John yu

POLYMER LATICES AND THEIR APPLICATIONS

POLYMER LATICES
AND THEIR APPLICATIONS

Edited by

K. O. CALVERT

B.Sc., F.P.R.I., C.Chem., M.R.S.C.

Projects Manager, Materials Technology, Dunlop Ltd, Birmingham, UK

MACMILLAN PUBLISHING CO., INC.
NEW YORK
COLLIER MACMILLAN CANADA, LTD.
TORONTO

Published in the USA by
SCIENTIFIC & TECHNICAL BOOKS
MACMILLAN PUBLISHING CO., INC.
866 Third Avenue, New York, N.Y. 10022

Distributed in Canada by
COLLIER MACMILLAN CANADA, LTD.

British Library Cataloging in Publication Data

Polymer latices and their applications.
 1. Rubber, Artificial
 I. Calvert, K. O.
 678'.71   TS1925

Library of Congress Catalog Card Number 81-84513
ISBN 0-02-949280-7

WITH 35 TABLES AND 22 ILLUSTRATIONS

© APPLIED SCIENCE PUBLISHERS LTD 1982

登記證：內版台業字第 1423 號
發行人：楊　　明　　德
　　　　台北市襄陽路 13～2 號
發行所：高立圖書有限公司
　　　　台北市襄陽路 13～2 號
　　　　電　話：3 6 1 5 3 3 0 號
　　　　郵政劃撥帳戶 105614 號
印刷所：東雅印製廠有限公司
　　　　台北市西藏路 528～530 號
中華民國 71 年　月　日

PREFACE

The aim of this book is to present a concise, technological account of the international polymer latex industry, including major developments up to 1980. Unlike most earlier books on latex, it is written by industrial experts in the various fields of latex technology, and assumes only a basic appreciation of latex science. It seeks to gather together in a readily accessible form a wealth of up-to-date practical experience, many aspects of which have not been published previously.

The book does not pretend to be a literature review and does not seek to describe historical events in any detail. It does, however, attempt to treat the subject matter from an international viewpoint, although the authors are all from the United Kingdom.

The book is not meant to compete with Dr Blackley's excellent book 'High Polymer Latices' which was published in 1966. Dr Blackley's volume is unreservedly recommended to readers who desire a more academic and detailed treatment of the subject.

The first four chapters of this book are devoted to latices themselves and the other seven chapters describe commercial applications of latex. The first chapter introduces the types of latex that are available, their properties and uses. It also comments on the handling and storage of latices. Commercial types of natural rubber latex, their principal properties and applications are described in the second chapter. The following chapter deals with the many types of synthetic latex that are available, covering the whole rubber to resin range, with their main properties and uses. The fourth chapter describes rubber latex specifications and codification and test methods for rubber and resin latices, primarily from the international viewpoint.

The last seven chapters of the book are devoted to the major applications of polymer latices. Processes are described, with formulations and product properties, of latex uses in carpets, synthetic binders, water-based adhesives and paints, latex dipping, moulded latex foam and finally diverse latex applications. The chapter on carpets covers latex anchor coats, secondary backings and gel and non-gel foam backings. That on synthetic binders embraces non-woven fabrics, papers and woven textiles while that on adhesives concentrates on the footwear, building, packaging and wood industries. The next chapter describes water-based decorative paints and that on latex dipping relates to the production of rubber gloves, prophylactics and medical articles. The Dunlop and Talalay processes are discussed in the latex foam chapter and, lastly, diverse latex applications embrace latex thread, rubberised hair and coir, and other uses.

As far as possible, trade names have been avoided in the writing of this book. On the other hand, the terminology of the International Standards Organization has been adopted wherever relevant and physical quantities have been expressed throughout in SI units.

Acknowledgements must first be directed to the co-authors of this book for their co-operation and fine contributions. In addition, I would particularly like to thank my colleagues H. M. Collier and R. J. Prichards for their valued assistance in the writing and editing of this volume and Dunlop Limited for permitting me to undertake this interesting task.

K. O. CALVERT

CONTENTS

Preface	v
List of Contributors	ix
1. Introduction	1
K. O. CALVERT	
2. Natural Rubber Latices	11
K. O. CALVERT	
3. Synthetic Latices	21
A. A. J. FEAST	
4. Latex Specifications and Test Methods	47
K. O. CALVERT	
5. Latex Applications in Carpets	71
D. PORTER	
6. Synthetic Latex Binders	93
R. CHARLTON, J. T. BOWDEN and B. SMITH	
7. Latex Adhesives	119
B. A. WOODLEY, J. PRITCHARD and A. A. ARMSTRONG	
8. Latex Paints	145
K. SELLARS and G. R. BROWN	
9. Latex Dipping	173
D. M. BRATBY	
10. Moulded Latex Foam	207
J. C. FALLOWS	
11. Diverse Latex Applications	229
T. D. PENDLE and A. D. T. GORTON	
Index	255

LIST OF CONTRIBUTORS

A. A. Armstrong

  Assistant Development Manager — Latex, Dunlop Semtex Ltd, Industrial Products Division, Chester Road, Erdington, Birmingham B35 7AL, UK.

J. T. Bowden

  Technical Department, Latex Division, Revertex Ltd, Temple Fields, Harlow, Essex CM20 2AH, UK.

D. M. Bratby

  Research and Development Manager, LRC Products Ltd, North Circular Road, London E4 8QA, UK.

G. R. Brown

  Applications Development Manager, Harlow Chemical Company Ltd, Temple Fields, Harlow, Essex CM20 2AH, UK.

K. O. Calvert

  Projects Manager, Materials Technology, Dunlop Ltd, Technology Division, Kingsbury Road, Birmingham B24 9QU, UK.

R. Charlton

  Technical Department, Latex Division, Revertex Ltd, Temple Fields, Harlow, Essex CM20 2AH, UK.

LIST OF CONTRIBUTORS

J. C. FALLOWS

Development Manager, Dunlop Ltd, Dunlopillo Division, Coronation Road, Cressex Industrial Estate, High Wycombe, Bucks HP12 3SB, UK.

A. A. J. FEAST

Technical Control and Development Manager, The International Synthetic Rubber Company Ltd, Brunswick House, Brunswick Place, Southampton SO9 3AT, UK.

A. D. T. GORTON

Principal Technologist, Tun Abdul Razak Laboratory, The Malaysian Rubber Producers' Research Association, Brickendonbury, Herts SG13 8NL, UK.

T. D. PENDLE

Senior Principal Technologist and Latex Group Leader, Tun Abdul Razak Laboratory, The Malaysian Rubber Producers' Research Association, Brickendonbury, Herts SG13 8NL, UK.

D. PORTER

Divisional Technical and Development Director, Lintafoam (Manchester) Ltd, Broadway, Haslingden, Rossendale, Lancashire BB4 4LS, UK.

J. PRITCHARD

Technical Manager, Dunlop Semtex Ltd, Industrial Products Division, Chester Road, Erdington, Birmingham B35 7AL, UK.

K. SELLARS

Technical Manager, Harlow Chemical Company Ltd, Temple Fields, Harlow, Essex CM20 2AH, UK.

B. SMITH

Technical Department, Latex Division, Revertex Ltd, Temple Fields, Harlow, Essex CM20 2AH, UK.

B. A. WOODLEY

> Customer Development Manager, Dunlop Semtex Ltd, Industrial Products Division, Chester Road, Erdington, Birmingham B35 7AL, UK.

Chapter 1

INTRODUCTION

K. O. CALVERT
Dunlop Ltd, Birmingham, UK

DEFINITION OF LATEX

A polymer latex is a colloidal dispersion of a rubber or a plastics material in an aqueous medium. The polymeric material may be a polymer of a single, small, ethylenically unsaturated organic compound or a copolymer of two or more such compounds. Usually, the polymer particles in the latex have a diameter of less than 1 μm, although a few latices do contain a small proportion of larger particles.

The stability of a polymer latex is due basically to the presence of surface-active material at the interface between the polymer particle and the aqueous phase. The majority of latices are anionic in character because their polymer particles carry a negative charge.

By the above definition, non-aqueous dispersions of polymeric materials, such as polyvinyl chloride plastisols, are outside the scope of this book.

TYPES OF LATEX

A wide range of polymer latices is available commercially. The principal varieties contain the following polymeric materials:

- Natural rubber
- Styrene–butadiene copolymers
- Acrylonitrile–butadiene (nitrile) rubbers
- Chloroprene rubbers
- Acrylic (co)polymers

Vinyl acetate (co)polymers
Vinyl chloride (co)polymers
Synthetic cis-polyisoprene
Butyl rubber

Natural rubber latices are concentrated and purified forms of the latex obtained from the botanical source *Hevea brasiliensis* which is cultivated in tropical regions. The second and largest group of polymer latices (shown above) consists of the many varieties of synthetic latex that are produced by emulsion polymerisation. The third group comprises latices which are made by solution polymerisation and conversion of the polymer solution to a latex.

The principal varieties of polymer latex listed above may in turn be divided into a number of types. For example, as described in the next chapter, there are several recognised types of natural rubber latex.

The many types of styrene–butadiene latex range in polymerised styrene content from zero, in polybutadiene latex, to 100 per cent in polystyrene latex. They include reinforced styrene–butadiene rubber latices and styrene butadiene terpolymer latices in which the third monomer is a carboxylic compound or vinyl pyridine. The nitrile latices do not have such a wide spread of polymerised acrylonitrile content but they also include carboxylic types.

In principle, any polymeric material that is produced by solution polymerisation can be converted to a latex. In practice, however, the cost of conversion of polymer solution to latex greatly limits the number of latices in this group that are commercially available. Synthetic latices are described, in much more detail, in Chapter 3.

PROPERTIES OF LATEX

The majority of commercial polymer latices have a total solids content of at least 40 per cent, and a few have a polymer content of nearly 70 per cent. They contain small quantities of non-polymer solids, such as surface-active agents, the amount approaching 10 per cent (calculated on the polymer) in the case of some synthetic latices of small particle size. Most types of polymer latex are mildly alkaline, with a pH in the vicinity of 10. They differ considerably in viscosity, depending on their total solids content and particle size distribution.

The most outstanding property of natural rubber latex is its high wet-gel strength which is especially beneficial in the manufacture of dipped or unsupported products. Many synthetic latices are tailored to particular applications and, therefore, have specific properties of, for example, adhesion or solvent resistance.

APPLICATIONS OF LATEX

In general, a latex is more expensive on a dry basis than its dry polymer counterpart. However, since it is a liquid, it is usually easier to process. Furthermore, a number of processes, such as latex backing of carpets, can be readily automated. Latex polymers also tend to have better physical properties and to be less prone to contamination. However, use of polymer latices is restricted to the manufacture of thin articles or components (up to $c.$ 2 mm thick), or thicker products with an inter-communicating cellular structure, because of the need to remove water in the latter stages of the process. The cost of water removal, and the attendant effects of shrinkage, are important, and sometimes limiting factors, in latex processes.

Commercial applications of polymer latices are most varied, as will be clear from the contents of later chapters. In the carpet industry, considerable quantities of latex are used for anchor coats, secondary backings and gel and no-gel foam backings. Other major latex applications are in binders for non-woven fabrics, papers and woven textiles. There are also a multitude of latex adhesives formulated for the footwear, building, packaging and wood industries. Water-based paints is another application area that consumes large amounts of latex. Dipping processes with latex for the manufacture of rubber gloves and similar articles are important too, as is moulded latex foam, which was the first major commercial application of latex. There are also many other latex applications, of which latex thread and rubberised hair are examples.

World consumption of all types of latex in 1980 was expected to be in excess of 2 million dry tonnes. In 1978 consumption of rubber latices alone exceeded 700 000 dry tonnes, of which nearly 300 000 dry tonnes was natural rubber latex. The percentage usage of natural rubber latex varies considerably from country to country, with a general tendency for it to be higher in less industrially developed regions.

HANDLING OF LATEX

Handling and storage of latex should be designed to avoid the following situations:

    (i) contact with unsuitable materials,
    (ii) all forms of contamination,
    (iii) high shear conditions,
    (iv) extremes of temperature,
    (v) stratification (creaming) of the latex,
    (vi) aeration of the latex,
    (vii) dehydration (skinning) of the latex surface, and
    (viii) loss of volatile preservatives (if any) from the latex.

Latex should not be allowed to come into contact with an absorbent or permeable surface or any material with which it (usually its aqueous phase) can react. Particularly unsuitable materials are mild steel, copper and its alloys. Examples of recommended contact materials are stainless steel (with a polished surface), plain steel coated with a phenolic or epoxy resin (that preferably has been baked on) or a glass lining that is resistant to high pH levels. Certain bitumastic or paraffin wax coatings are also appropriate with some latices. Latex is normally supplied to the consumer in drums or from road tankers or rail tank cars that have smooth contact surfaces that are chemically resistant to the latex.

Equipment used for handling or storage should be kept clean, to minimise the risk of contamination of the latex. Steps need also to be taken to prevent contamination from external sources by having a close-fitting cover on the latex container. The most dangerous potential contaminants are solutions (usually acidic) or solvents that destabilise the latex. Even water can have a destabilising effect if it has a high mineral content. Exposure to acidic gases is also to be avoided and unnecessary contact with air, which contains carbon dioxide, should be prevented. Similarly, latex ought not to be stored or used in the proximity of any equipment that generates carbon dioxide, for example, certain types of gas or oil heaters.

Latex should not be subjected to high shear conditions because of its adverse effect on stability. Positive displacement gear and piston pumps and closed-impeller centrifugal pumps are not advised for use with latex. Recommended types of pump are the open-impeller, low speed, centrifugal varieties and diaphragm pumps. Alternative means of transferring latex, that are sometimes applicable, are the use of gravity or the pressure system. In the latter, the latex for transfer must be contained in a pressure-tested vessel, and it is important that only clean compressed air is applied to it and the air pressure is released before aeration of the latex occurs.

Latex ought not to be subjected to extremes of temperature. Its temperature should be prevented from falling below 5°C (unless it is a freeze–thaw stable grade of latex) or from exceeding 40°C and preferably should be maintained between 10 and 30°C. With decreasing temperature, the viscosity of latex increases, making handling more difficult, until the latex freezes at a few degrees below 0°C. At excessively high temperatures, skinning of the latex surface is accelerated and creaming tendencies are encouraged, unless the latex container is tightly closed and the surface of the latex is regularly disturbed. Ideally, vessels, pipe lines and fittings for latex are insulated or jacketed to counteract the effects of extremes of temperature.

Latex tends to cream (stratify) on standing, at a rate that depends on its concentration, its particle size and distribution, its viscosity, the geometry of its container and the temperature. Creaming is best prevented by low speed mechanical agitation that is sufficient to gently disturb the latex surface without causing aeration. Unnecessary stirring is to be avoided and this objective can often be achieved by adoption of periodic, instead of continual, agitation. Alternatively to mechanical stirring, latex can be homogenised by transferring it from one container to another in such a manner that aeration is not encouraged.

Aeration of latex is undesirable because it makes transfer operations more difficult, it hastens dehydration of the latex surface and it exposes the latex to the destabilising influence of carbon dioxide. Furthermore, the presence of air, which is not easily removed from latex, is troublesome in several latex processes. Aeration is caused by too vigorous mechanical agitation, by cascading latex from one receptacle to another or by applying excessive air pressure to the latex.

Skinning or dehydration of the latex surface is a wasteful situation that arises from exposure of dormant latex to unsaturated air. It is accelerated by foam on the surface of the latex, by elevated temperature or by draughts. Skinning can be minimised by mechanical agitation and by having a tight-fitting cover on the latex vessel.

Some latices, particularly natural rubber latex, contain volatile preservatives such as ammonia, reduction of which may have a deleterious effect on quality. Loss of volatiles can be limited by use of well sealed containers, lowering of temperature and prevention of aeration of the latex.

Although latices are not dangerous materials unless they are ingested, sensible safety precautions should be taken when handling them. Approved eye protectors ought always to be worn and protective gloves or a barrier skin cream are advised. When spraying latex, effective procedures should be adopted to avoid inhalation of spray droplets.

A vessel that has held latex containing a volatile ingredient must not be entered until it has been established that the concentration of that material is below its threshold limiting value throughout the height of the vessel. If necessary, the required condition can be achieved by forced ventilation of the vessel. Normal safety measures should also be observed by equipping the person inside the vessel with a lifeline and stationing another person outside the vessel in direct contact with him.

RANGE OF COMPOUNDING INGREDIENTS

Only a few specialised latices can be used directly as supplied by the manufacturer. The majority of latices require the addition of compounding ingredients before they can be converted into a finished product or component.

The range of compounding ingredients used in latex technology is extremely broad, and wider than that for solid polymers. Compounding ingredients for latex may be divided into the following categories:

(a) stabilisers including surfactants,
(b) vulcanising agents,
(c) vulcanisation accelerators,
(d) antioxidants,
(e) fillers,
(f) viscosity modifiers (thickeners),
(g) gel sensitisers.

Latex stabilisers comprise alkalis, protective colloids and numerous surface-active agents. Alkalis may be fugitive, such as ammonia, or permanent like potassium hydroxide. Various protective colloids, casein for example, are still employed but their use is diminishing with the application of synthetic surfactants that are similarly effective and more consistent in stabilising power. Surface-active agents include carboxylates such as potassium oleate, sulphonates, organic sulphates, cationic surfactants, amphoteric and non-ionic materials. A widely used sulphonate is the powerful dispersing agent sodium naphthalene formaldehyde sulphonate. A typical example of a straight-chain alkyl sulphate is sodium dodecyl (lauryl) sulphate. Examples of cationic and amphoteric materials are, respectively, lauryl pyridinium chloride and C-cetyl betaine. Many of the non-ionic stabilisers used in latex technology are condensation products of ethylene oxide with fatty acids, fatty alcohols or phenols; for instance the condensate of 1 mole of oleyl alcohol with 25 moles of ethylene oxide.

The most common vulcanising agent is elemental sulphur. Sulphur donors such as tetramethyl thiuram disulphide are also used, to produce in particular a 'sulphurless' vulcanisate with enhanced heat resistance. Metal oxides, including zinc oxide, are sometimes employed as cross-linking agents with carboxylic latices.

Of the vulcanisation accelerators for latex, the most important are the dialkyl dithiocarbamates and thiazoles. Zinc diethyl dithiocarbamate (ZDEC) is widely used and mercaptobenzthiazole (MBT), often as its zinc salt (ZMBT), is a quite common secondary accelerator. Various amines, such as diphenyl guanidine (DPG), find application usually as vulcanisation activators. Zinc oxide also has an activating function.

Latex antioxidants fall into three main classes, namely amine derivatives, hindered phenols and styrenated phenols. Amine antioxidants protect the latex polymer best against the effects of heat and trace metals but they tend to cause discoloration during ageing. Phenolic antioxidants are not so effective, except against light, but they have better colour properties. Antiozonants are also used occasionally and sometimes a wax is included in a latex formulation to serve a similar purpose.

The main types of fillers used in latex technology are the clays (hydrated aluminium silicates), calcium carbonates and alumina trihydrates. Larger particle-size (and cheaper) fillers are employed where high loadings are required. Alumina trihydrates are usually more expensive but they have gained in popularity in recent years because of the beneficial effects that they have in certain product flammability tests.

There are several types of latex thickeners but polyacrylates, celluloses and alginates are probably the most important. Caseinates and colloidal clays are on the decline because their effects are less consistent than those of the modern synthetic or more refined natural products.

Gel sensitisers may be divided into two main groups, namely those with a delayed-action effect (at ordinary temperatures) and those that are activated by elevation of temperature. The best known delayed-action gelling agent is sodium silicofluoride which was employed in the first large-scale latex foam process. A common heat sensitiser is polyvinyl methyl ether, but ammonium salts (in conjunction with zinc oxide) are still used in certain circumstances.

In addition to the above compounding ingredients, there are several others that are specific to a type of latex or to a particular application. Some of them are indicated in subsequent chapters of this book.

PREPARATION OF COMPOUNDING INGREDIENTS

In most cases, it is necessary to distribute the compounding ingredient uniformly in an aqueous medium before adding it to latex. The procedure required depends on whether the material, solid or liquid, is soluble or insoluble in water.

A water-soluble ingredient can be added to latex as an aqueous solution, but there are practical limitations. The solution must not be so concentrated that it has a shock effect on the latex and its pH should, if possible, be similar to that of the latex for the same reason. A solution cannot be used if it contains destabilising (e.g. polyvalent) ions, unless the latex has been adequately stabilised beforehand.

Most latex compounding ingredients are water-insoluble solids. In general they can be dispersed in water with the aid of a dispersing agent like sodium naphthalene formaldehyde sulphonate. The concentration of ingredient in the dispersion is between 30 and 70 per cent, and 50 per cent is a common choice. The water used should be demineralised, distilled or soft, and it should contain an alkali (or acid) to make the pH of the dispersion similar to that of the latex. A typical dispersion formulation, in parts by mass, is the following:

Solid compounding ingredient	50
Dispersing agent	1·5
Water	48·5

The crude dispersion has usually to be milled to reduce its particle size to the colloidal range (normally below 5 μm and for some applications to less than 2 μm). This is done in a ball mill, attritor or colloid mill. The finished dispersion should be stored in a container made of a material that is chemically resistant to it. It should be agitated, preferably continually and slowly, to prevent sedimentation of the compounding ingredient.

Liquid compounding ingredients that are water-immiscible can be emulsified in water with the aid of an emulsifying agent such as potassium oleate. The emulsifier may be preformed or produced *in situ*. An example of the latter type of formulation, in parts by mass, is as follows:

Liquid compounding ingredient	50
Oleic acid	1·5
Potassium hydroxide	0·3
Water	48·2

The blend of compounding ingredient and oleic acid is added slowly to the aqueous alkali solution, with agitation, and the mixture is sheared in a homogeniser or colloid mill. The finished emulsion should be stored in similar manner to a dispersion.

In a few cases, compounding ingredients can be added directly to latex, without previous dispersion or emulsification. Prior stabilisation of the latex is usually required and this mode of compounding ought not to be undertaken without adequate pilot trials. Large particle-size fillers and mineral oils are examples of compounding ingredients that have been added directly to high solids styrene–butadiene latices.

COMPOUNDING OF LATEX

Although most latex compounds involve only one latex, there are formulations that use a blend of different types. A case in point is the combination of natural rubber latex and high solids styrene–butadiene latex, that is employed in some foam rubber manufacture.

Blending of different types of latex can present difficulties unless their polymer/water interfacial systems are similar. If the interfacial systems are dissimilar, as is often the situation, thickening usually occurs on mixing the latices. This thickening can be pronounced and, although generally it is only temporary, several hours may elapse before the viscosity of the latex blend is low enough for subsequent compounding. A more practical approach is to add a suitable stabiliser to one of the latices, preferably that with the greater soap deficiency, before blending it with the other latex. In this way thickening can be prevented or substantially reduced. With the particular latex blend mentioned above, thickening is greatly diminished if potassium oleate (1 phr) is added to the natural latex before it is mixed with the styrene–butadiene latex.

The order in which compounding ingredients are incorporated into latex is not usually critical, provided any required stabilisers are added first and any thickeners or destabilising (sensitising) ingredients are introduced last. With more complex formulations, it is not unusual for the latex (or latex blend) to be compounded in two or more stages, to give better control of maturation and processing.

Compounding of latex should be performed in a vessel that is chemically resistant to the latex and the compounding ingredients. Preferably the vessel should be equipped with a mechanical stirrer, the blades of which are located low in the vessel and have a shape that induces effective agitation without foaming. The compounding ingredient is best added to the continually stirred latex in a slow steady stream, with a minimum of splashing.

The compounded latex should be stored in a covered, chemically resistant vessel that has a temperature-controlled jacket to regulate maturation and viscosity characteristics. It is advisable to stir the compounded latex continually, at low speed, particularly if it contains high-density ingredients, such as zinc oxide or fillers.

Latex formulations are usually calculated on dry rubber content (DRC) in the case of natural rubber latex and on total solids content with synthetic latices. However, it is preferable to compound high solids synthetic rubber latices on dry polymer content (Chapter 4). In calculating the formulation, the concentration by mass of the compounding ingredient itself, rather than the total solids content of its dispersion, should be employed.

An illustration of a latex compound, in parts by mass, is given below.

Ingredient	Parts of principal	Wet parts
Natural rubber latex (60 per cent DRC)	100	166·7
20 per cent potassium hydroxide solution	0·5	2·5
50 per cent sulphur dispersion	0·5	1·0
50 per cent antioxidant dispersion	1·0	2·0
50 per cent zinc diethyl dithiocarbamate dispersion	0·5	1·0
50 per cent zinc oxide dispersion	1·0	2·0

Several technological formulations for latex compounds, based on various types of latex, are given in the chapters of this book that are devoted to latex applications.

Chapter 2

NATURAL RUBBER LATICES

K. O. CALVERT
Dunlop Ltd, Birmingham, UK

Natural rubber latex as it emerges from the tree (field latex) has a dry rubber content of only about 30 per cent. It can be preserved by the addition of ammonia or caustic alkali, but its low rubber content and high non-rubber solids content severely limits the usefulness of preserved field latex.

Latex concentration processes were established in the decade around 1930 and they made available the products on which latex technology was built. Three distinct processes were developed for increasing the dry rubber content of natural latex to 60 per cent or more. The first was evaporation, in which alkali and soap were added to field latex before it was concentrated by heat. The second process was centrifugation of field latex, preserved with ammonia, in a machine with a plurality of conical separator discs. The other process was creaming, in which a creaming agent such as ammonium alginate was added to ammonia-preserved field latex and the resultant upper cream layer was separated from the underlying serum fraction.

It should be noted that evaporated natural latices contain all the non-rubber solids present in field latex. On the other hand, centrifuged and creamed concentrates contain only about 30 per cent of the non-rubbers of field latex, because of the separative effects of these methods of concentration. Latex can, however, be concentrated to a greater extent by evaporation than by the other two processes.

Mention should also be made of a fourth method of latex concentration, namely electrodecantation. This type of natural rubber latex was produced commercially for a few years up to 1959.

Of the three surviving types of concentrated natural rubber latex, nearly 90 per cent is of the centrifuged variety and the remainder is divided, roughly equally, between the creamed and evaporated types.

CENTRIFUGED NATURAL LATICES

Centrifuged natural latices are available commercially as high-ammonia (0·6–0·8 per cent ammonia) and low-ammonia (0·2 per cent ammonia) types. The former is preserved solely, or principally, with ammonia whereas the latter contains at least one other preservative in addition to ammonia.

High-ammonia latex was the original type but it is being replaced steadily by the low-ammonia varieties, a trend that is likely to continue. The reasons for this trend are the increasing resistance of latex bacteria to ammonia alone, the fact that a number of latex manufacturing processes require a low alkalinity and the recent introduction of stringent regulations governing ammonia concentrations in factory atmospheres.

Three types of low-ammonia latex had been established commercially by 1960. They contained, in addition to 0·2 per cent ammonia, the following secondary preservatives:

(i) 0·2 per cent sodium pentachlorphenate (LAPCP latex),
(ii) 0·25 per cent boric acid and 0·05 per cent sodium pentachlorphenate (LABA latex), and
(iii) up to 0·1 per cent zinc diethyl dithiocarbamate (LAZDC latex).

The main characteristics of these three types are high stability with LAPCP latex, less effective long-term preservation with LABA latex and a lower level of stability with LAZDC latex. Sodium pentachlorphenate is also, unfortunately, a material that is deprecated on environmental grounds and latex containing it is prohibited in some countries. LABA latex is similarly subject to certain national restrictions.

The commercial introduction of two more, closely related, varieties of low-ammonia latex in about 1976 has accelerated the movement to this type of latex. These two varieties contain the following secondary preservatives:

(i) less than 0·05 per cent tetramethyl thiuram disulphide (TMTD) and 0·02–0·03 per cent zinc oxide, and
(ii) less than 0·05 per cent sodium diethyl dithiocarbamate (SDEC) and $c.$0·03 per cent zinc oxide.

The advantages of the TMTD/ZnO and SDEC/ZnO types of low-ammonia latex are that the secondary preservatives are materials that are widely used in latex technology and their concentrations are lower than those of the preservatives in the earlier types. The latices are known commercially as LATZ, LAZN or LATD latex, qualified usually by the manufacturer's name. Their properties are similar to those of high-ammonia latex of good quality, except for alkalinity and pH and a greater stability towards added zinc compounds.

High-ammonia latices that are preserved solely with ammonia are defined by the International Standards Organization as HA latices. High-ammonia latices that contain preservatives (other than formaldehyde) in addition to ammonia are defined, by the same organization, as XA latices. Both of these types are commercially available, although the latter variety has become more common in recent years.

PROPERTIES OF LOW-AMMONIA (CENTRIFUGED) LATEX

Properties of a typical commercial low-ammonia latex of the TMTD/ZnO type are as follows:

Total solids content, per cent	61·5
Dry rubber content, per cent	60·0
Alkalinity, per cent ammonia	0·20
Mechanical stability, s	1000
Zinc stability time (ZST), s	500
Potassium hydroxide number	0·5–0·6
Volatile fatty acid number	0·02–0·03
Carbon dioxide number (see Chapter 4)	0·15
Viscosity (Brookfield L) at 60 rpm, mPa.s	70
Surface tension, mN/m	38
pH	9·7

The difference between total solids content and dry rubber content is typical of a once-centrifuged natural latex. Alkalinity, which is determined by acidimetry, is a more reliable property of natural latex than pH.

Mechanical stability is an important property of natural rubber latex. It is controlled by the latex manufacturer by the addition of ammonium laurate in amounts that are much less than the quantity of soap that is formed naturally in the latex by hydrolysis of phospholipids. Control of mechanical stability is a complex matter since this property changes with the age of the latex, particularly at tropical temperatures. Commercial natural latex is, however, substantially constant in mechanical stability once it is in storage in temperate countries.

The zinc stability of latex of the TMTD/ZnO type, as measured by the ZST test, is high although it is less than that of LAPCP latex; it is greater than with most varieties of high-ammonia latex.

The good state of preservation in TMTD/ZnO latex is indicated by its low volatile fatty acid and carbon dioxide numbers. The potassium hydroxide number is not such a reliable guide to quality since it measures all the anions that are combined with ammonia in the latex. Most of the anions, comprising the more prevalent non-volatile acid radicals and the amino-acids as well as volatile fatty acids and carbonates/bicarbonates, have a destabilising influence but the higher fatty acids have an opposite, stabilising, effect. All these anions are titrated collectively in the potassium hydroxide number determination.

The viscosity of TMTD/ZnO latex is similar to LAPCP latex and slightly lower than that of high-ammonia latex. Its surface tension represents an apparent soap deficiency of approximately 0·6 per cent on total solids, as determined by titration with potassium oleate. This deficiency is not, however, cause for concern since the particles in natural latex, unlike synthetic latices, are stabilised with proteins as well as soaps.

Of the metallic elements present in natural latex, the most important are potassium, magnesium, iron and copper, together with any zinc that is added during the manufacturing process. Natural latex produced in Malaysia contains the following approximate amounts of these elements, in milligrams per kilogram of latex:

Potassium	1100
Magnesium	20
Iron	3
Copper	2

Potassium of botanical origin is in the largest concentration. The magnesium content of field latex varies considerably with the clonal variety of tree and it is reduced by the latex manufacturer by controlled precipitation with diammonium hydrogen phosphate prior to centrifuging. The potentially deleterious element copper is present only in trace amounts and iron content is of a similar magnitude.

Two other useful properties of natural rubber latex are its density and specific heat capacity. For most technological purposes, the density of latex of 60·0 per cent dry rubber content may be taken as 0·945 Mg/m^3 (g/cm^3) at 20°C, reducing linearly to 0·940 Mg/m^3 at 30°C. For costing purposes, however, it is customary to determine the density of the latex. The specific heat capacity of natural latex of 60·0 per cent dry rubber content, calculated from the values for its constituents, is approximately 2·75 kJ/kg °C. Thus if, for example, it is required to increase the temperature of 100 Mg (100 tonnes) of latex by 5°C, nearly 1400 MJ of heat needs to be put into the latex.

NON-RUBBER SOLIDS

The principal non-rubber solids in well-preserved natural latex are proteinaceous material, higher fatty acid soaps, salts of non-volatile acids (NVA) and carbonates/bicarbonates (CO_3). The first two of these components are stabilisers, whereas the others have the opposite effect.

Concentrations of the main non-rubber solids in commercial latex of the low-ammonia TMTD/ZnO type are approximately as follows (as percentages of latex total solids):

Protein	1·65
Soap	0·80
Non-volatile acids	0·45
Carbonate (CO_3)	0·08

Two-thirds of the protein is associated with the rubber phase of the latex. This fraction is acid-coagulable and appears in the dry rubber content. Most of the soap content is derived from alkaline hydrolysis of naturally occurring phospholipids. The remainder is the small proportion that is added to facilitate centrifuging and adjust mechanical stability.

The non-volatile acids (NVA) consist, in the main, of citric and malic acids, phosphate and sulphate. Various other non-volatile acids and amino-acids are also present in latex in smaller concentrations.

The difference between total solids content and dry rubber content increases slightly after the latex has left its country of manufacture. The reason for this change is the small decrease (approximately 0·05 per cent) in dry rubber content that occurs without alteration of total solids content. The reduction in dry rubber content is due to delayed hydrolysis and loss of a minor proportion of the non-rubbers from the rubber phase to the aqueous phase.

TWICE-CENTRIFUGED LATEX

In some latex applications, it is advantageous to use a latex with a lower non-rubber solids content. Such a latex contains less hygroscopic material, which facilitates washing and drying of the product made from it, and gives certain improvements in product properties. It is, however, more expensive than the standard types of latex.

Reduction of non-rubber solids can be achieved by centrifuging, or creaming, a latex that has already been concentrated once. The sole commercial latex of this type (in 1980) is twice-centrifuged latex, which is produced only in small volume, usually to meet a specific request from a consumer. It is used for the manufacture of certain medical articles and products requiring improved resilience properties.

Twice-centrifuged latex is prepared by diluting once-centrifuged latex to 30 per cent dry rubber content with water containing ammonia and then recentrifuging to 60 per cent dry rubber content. This procedure is modified by some manufacturers and one variant is known as 'substage' latex. Twice-centrifuged latex can be either high- or low-ammonia.

Typical properties of a twice-centrifuged latex of the low-ammonia TMTD/ZnO type are as follows:

Total solids content, per cent	60·6
Dry rubber content, per cent	60·0
Alkalinity, per cent ammonia	0·20
Potassium hydroxide number	0·25
Volatile fatty acid number	0·01
Carbon dioxide number	0·06
Potassium, mg/kg	400
Protein, per cent of total solids	1·0
Soap, per cent of total solids	0·7
Non-volatile acids, per cent of total solids	0·15

The difference between total solids content and dry rubber content is less than half the level in once-centrifuged latex. On the other hand, more than 60 per cent of the stabilising proteins and soaps are retained after the second centrifuging, indicating that most of these components are associated with the rubber phase. Destabilising constituents — non-volatile acid (NVA) radicals, carbonates/bicarbonates and volatile fatty acids — are decreased to roughly one-third by recentrifuging, in accordance with aqueous phase proportionment. The potassium concentration is lowered similarly. The potassium hydroxide number is not reduced to the same extent by the second centrifuging because little of its soap component is lost to the separated serum fraction.

CREAMED NATURAL LATEX

Creamed natural latex has a greater concentration, higher viscosity and, it is claimed, better filterability than centrifuged latex. It may be either high- or low-ammonia, although only the former variety is in regular commercial production. It is more expensive than (once-) centrifuged latex.

Properties of the high-ammonia type of creamed natural latex approximate to the following:

Total solids content, per cent	68·0
Dry rubber content, per cent	66·8
Alkalinity, per cent ammonia	0·65
Mechanical stability, s	1600
Zinc stability time (ZST), s	175
Potassium hydroxide number	0·62
Volatile fatty acid number	0·07
Viscosity (Brookfield R) at 20 rmp, mPa.s	1100

Total solids content and dry rubber content are substantially greater than with centrifuged latex, and the difference between them is slightly less. Viscosity is relatively high as supplied but it is similar to that of centrifuged latex at comparable total solids. Stability to added zinc, as measured by the ZST test, is on the low side.

Creamed natural latex is used in the manufacture of extruded thread and in other latex applications in which a higher concentration and viscosity are advantageous.

EVAPORATED NATURAL LATEX

There are three types of evaporated natural rubber latex, two of which have a greater total solids content than centrifuged latex. One type is stabilised with potassium hydroxide and soap and has a total solids content of 73 per cent. The second type, which was developed from the first, is the most important grade. It is stabilised with potassium hydroxide and ammonia and has a total solids content of 68 per cent. The third type of evaporated natural latex is stabilised with at least 0·6 per cent ammonia and has a total solids content of approximately 62 per cent.

All varieties of evaporated latex are characterised by a high level of non-rubber solids, amounting to up to 8 per cent of the latex. Since there is no separative stage in their manufacture, they contain the proportion of smaller rubber particles that is removed in the serum fraction in the centrifuging process. Consequently, they have a lower average particle size than the other types of natural latex and correspondingly a higher viscosity at comparable total solids. Evaporated latex of 73 per cent total solids content has the consistency of a paste and is supplied only in drums.

Because of its high stability, evaporated latex containing potassium hydroxide finds use in spreading and similar applications.

The alkalinity of evaporated latex stabilised with potassium hydroxide is at least 0·75 per cent expressed as the concentration of potassium hydroxide. Evaporated latices have a larger sludge content than the centrifuged and creamed types, as would be expected from their greater level of non-rubber solids.

SPECIALITY NATURAL LATICES

In addition to the types of natural latex that are in regular commercial supply, there are a few variants that can be produced if the additional cost premium can be justified. Examples of these speciality materials are freeze–thaw stable natural latex, methyl methacrylate grafted natural latex and centrifuged latex of high dry rubber content.

Natural latex can be made freeze–thaw stable by the addition of small quantities of sodium salicylate and laurate soap. These additions are best incorporated during manufacture of the latex. The treatment can be applied to either high-ammonia or low-ammonia latex.

The material obtained by grafting methyl methacrylate onto natural rubber in latex form is known as Heveaplus MG latex. The level of grafted monomer in the dry product is normally 30 per cent or 49 per cent. This type of latex has a reinforcing effect and, it is claimed, provides a means of substantially improving the tear and puncture strength of dipped latex articles.

Centrifuged latex can be produced with a dry rubber content that is as high as in creamed latex (67 per cent). Since at least two centrifuging stages are involved, high DRC centrifuged latex may be considered to be a variant of twice-centrifuged latex. Thrice-centrifuged natural latex has also been made on a very limited scale, but for scientific rather than technological purposes.

PREVULCANISED NATURAL LATEX

Natural rubber latex can be vulcanised in the latex state by the addition of a stabiliser, sulphur, an ultra-accelerator and zinc oxide, and heating to, and at, approximately 75°C for 2 to 3h. Excess vulcanising ingredients are minimised by the manufacturer by judicious compounding or by centrifuging the vulcanised latex or allowing it to sediment.

Most varieties of commercial, prevulcanised natural latex have a total solids content of about 60 per cent, an ammonia content of 0·6 per cent, a pH of 10·5 and a viscosity similar to that of natural latex itself. Low-, medium- and high-modulus grades are available. There is also a low-ammonia, medium-modulus grade of prevulcanised latex with an ammonia content of 0·3 per cent and a pH of 10·0.

Prevulcanised latex has the attraction that it does not require vulcanisation by the user. Drying of the latex article made from it is sufficient to achieve a vulcanised product with physical properties that are satisfactory for many purposes. Nevertheless, it should be borne in mind that a better balance of physical properties can often be obtained by post-vulcanisation of ordinarily compounded latex.

Prevulcanised natural latex is used for dipping, moulding and casting applications. The low-ammonia grade is particularly suitable for heat-sensitised dipping.

APPLICATIONS OF NATURAL RUBBER LATICES

Natural latex is used for or in the manufacture of many rubber products although it has been replaced, wholly or partially, by synthetic latices in a number of applications since about 1960. More often than not, this substitution resulted from cost rather than technical considerations.

Natural latex dominates the production of dipped goods and extruded thread and is widely used in water-based adhesives. It is also employed in the manufacture of moulded and cast articles, rubberised hair and coir and various other applications where its high strength is necessary.

Considerable quantities of natural latex are consumed in the production of moulded latex foam, particularly in the Far East. Carpet backing processes also employ a significant proportion of natural latex.

These applications, and others, of natural latex are described in Chapters 5, 7, 9, 10 and 11.

Chapter 3

SYNTHETIC LATICES

A. A. J. Feast
International Synthetic Rubber Co. Ltd, Southampton, UK

Synthetic latices may be defined as aqueous dispersions of polymer particles produced by the process known as emulsion polymerisation. Interest in these materials arose initially from the use of natural rubber latex and the prospect of making, by synthetic means, a latex similar to the naturally occurring product.

The earliest references to synthetic latices probably date back to pre-1920 with the work originating in Germany. It has been noted by several reviewers that all the essential elements of emulsion polymerisation were almost certainly known before World War I.

During the period 1930–35 emulsion polymerisation was being established as a method of producing synthetic rubber latices. Considerable advances were made during World War II in Germany and USA, when continuous processes for synthetic rubber production were developed for styrene–butadiene (SBR) and acrylonitrile–butadiene copolymers to alleviate the shortage of natural rubber. Since then, many other types have come onto the market with polyvinyl acetate and copolymers, acrylics, and carboxylic–SBR types being the major products.

Today, the annual consumption of synthetic latices of all types is estimated at about 800 000 dry tonnes for Western Europe and a further 900 000 dry tonnes for USA.

A breakdown of available figures of tonnages for various polymer types is given in Table 3.1.

EMULSION POLYMERISATION

Emulsion polymerisation has become very important industrially, and today large tonnages of latices and polymers are manufactured by this

TABLE 3.1
Synthetic latex consumption in Western Europe 1979–80

Type	'000 dry tonnes
Carboxylic–SBR	238
Vinyl acetate homo- and copolymer	237
Acrylics	150
High solids SBR	140
Acrylonitrile–butadiene copolymer	14
Polyvinylidene chloride	7–10
Polyvinyl chloride	7–9
Polychloroprene	5–8

process. Although it is not within the scope of this book to discuss the theoretical aspects in detail (several reviews[1,2,3] are available in the literature), it is considered that a general discussion is necessary so that the growth, use and scope of the process may be more readily understood.

Emulsion polymerisation has certain advantages over other industrial methods such as bulk or solution polymerisation, as follows:

(i) Very high molecular weight polymers may be obtained at fast reaction rates.

(ii) As invariably an aqueous medium is used, there are no problems with heat dissipation.

(iii) Use of water minimises cost as no recovery system is required as for an expensive solvent. Also there should be no particular fire or toxicity hazards.

(iv) Ease of control of each stage of the reaction, i.e. initiation, propagation, chain transfer and termination in conjunction with fairly low polymerisation temperatures ($<100°C$).

(v) The viscosity of a latex is independent of the polymer molecular weight (c.f. solution polymers), and high concentrations of low viscosity latex can be obtained, aiding agitation and heat transfer characteristics, and enabling material to be readily handled (pumped, etc.) at ambient temperature.

(vi) Polymer is obtained as a latex and is used as such in many end-use applications.

(vii) If desired, a solid product may readily be obtained after coagulation, washing and drying.

(viii) Emulsion processes are easily adaptable to continuous running conditions.

There are a number of disadvantages, but these are of a limited nature:

(i) In end-use applications, evaporation of water is a slow process because of its high latent heat.
(ii) Reactor capacity is not fully utilised as latices generally only contain up to a maximum of 55 per cent polymer solids as a result of direct polymerisation.
(iii) A solid polymer of high purity cannot be obtained as the isolated material will contain a certain proportion of non-polymer constituents. It must be mentioned, however, that the presence of insoluble soap derivatives is known to be advantageous in many applications.

An emulsion polymerisation will contain four basic ingredients, monomer, water, soap (dispersing agent, emulsifier or surfactant) and catalyst (initiator) in the following typical proportions:

		Parts by mass
Monomer(s)		100
Soap(s)	Up to	7
Initiator	Up to	1
Water	Up to	200

The properties of the latex formed and of the polymer or copolymer are very dependent on how the various constituents are put together.

A simplified, somewhat idealised, description of emulsion polymerisation is as follows:

(i) Emulsifier in aqueous solution above a certain concentration forms micelles (aggregates of molecules).
(ii) Addition of water insoluble monomer, which forms droplets stabilised by emulsifier. Some monomer solubilised in micelles.
(iii) Free radicals are generated by initiator and cause polymerisation inside the micelle.
(iv) Growing polymer particles obtain further monomer by diffusion from monomer droplets.
(v) Monomer droplets used up at about 60 per cent conversion and polymerisation continued until remaining monomer in polymer particles is consumed.

Soap (emulsifier)
This constituent could be considered as the most important as it must give a stable emulsion between monomer and water initially and during reaction, and give a stable latex finally.

There are four classes of soap, anionic, non-ionic, cationic and amphoteric, designated according to their hydrophilic groups. The most commonly used are anionic and non-ionic types. Anionic emulsifiers may be sodium or potassium salts and fatty and rosin acids, alkyl sulphates, sulphonates, sulphosuccinates or alkyl aryl sulphonates. Non-ionic emulsifiers are normally ethylene oxide condensates of long-chain alcohols or fatty acids. It it common practice for some applications to use mixtures of anionic and non-ionic emulsifiers.

Initiator
A large variety of initiator systems are available for use in emulsion polymerisation and those used industrially are based on the liberation of a free radical. Active free radicals may be produced by thermal decomposition or chemical interaction. Examples of free radical generation by thermal decomposition are given by organic and inorganic peroxides, i.e. lauryl peroxide, t-butyl hydroperoxide and ammonium peroxydisulphate. Radicals are produced by chemical interaction by using a peroxide or hydroperoxide in conjunction with a reducing agent, the so-called redox system.

Monomers
The most important monomers used commercially in emulsion polymerisation are styrene, butadiene, alkyl acrylates and methacrylates, vinyl acetate, acrylonitrile, chloroprene, vinyl chloride, vinylidene chloride, ethylene and 2-vinylpyridine, with all being found as constituents of co- or terpolymers and, with the exception of acrylonitrile and 2-vinylpyridine, as homopolymers.

Miscellaneous
Other important components of an emulsion polymerisation system are the modifier (chain transfer agent) which influences the molecular weight and molecular weight distribution of the polymer, and the electrolyte which can alter the soap efficiency and affect the rate of reaction, particle size and particle size distribution.

LATEX PROPERTIES

A list of latex properties is given below. Some properties apply to all latices, but others only apply to latices for use in specific end applications. The list is divided into general and specific properties.

General properties	*Specific properties*
Total solids content	Soap coverage
pH	Soap/rubber (resin) ratio
Viscosity	Chemical stability
Surface tension	Minimum film forming temperature
Particle size	
Particle size distribution	Glass transition temperature
Coagulum content	Freeze–thaw stability
Mechanical stability	Gel content (isolated polymer)
Residual monomer content	Mooney viscosity (isolated polymer)

METHOD OF PREPARATION

To obtain the required properties for any given latex, there are a large number of contributing factors that must be taken into account. These may be broadly divided into two groups, chemical and operational.

Chemical	*Operational*
Type and level of emulsifier	Reactor capacity and profile
Type and level of initiator	Agitation — stirrer type and speed
Type and level of modifier	Temperature, pH and pressure
Type and level of electrolyte	Mode of addition of emulsifier,
Phase ratio (monomers/water)	monomers, etc. (i.e. batch,
Relative water solubility of monomers	incremental or continuous)
Relative reactivities of monomers in copolymerisation	

Commercially, polymerisation techniques fall into three categories, namely batch, semi-continuous and continous.

Batch reactions
Most of the published experimental data and development of theories have resulted from the study of batch reactions. The recipe constituents are charged into a single stirred reactor and polymerisation initiated and

continued until the required conversion is attained. Reactor contents are discharged, reactor flushed out and the cycle repeated.

Reactions may be 'hot' ($>60°C$) or 'cold' ($5°C$) according to the monomers in use, and reactors are equipped with heating and cooling systems operating through a jacket or internal coils. Also, depending on the volatility of the monomers used, reactions may be carried out under pressure or under reflux conditions with a condenser in the system.

The type of agitation is very important and must be such that a good dispersion is maintained without excessive shear, which could lead to coagulum formation.

Batch reaction requires relatively simple equipment and, although not as efficient as other processes, is effective for production of lower volumes and speciality materials.

Semi-continuous reaction

This is probably the most widely used emulsion polymerisation technique. It is essentially a modification of the batch reaction process, and employs a single stirred reactor with monomer and possibly some other ingredients added continuously or incrementally over a certain period of reaction time. This technique has certain advantages in that it allows alteration of the polymer structure and is more efficient in reactor utilisation because the monomer can be added as fast as the heat of reaction can be removed. Also, controlled addition of a monomer emulsion can be used to control particle formation and stability. If free emulsifier is available during addition, so that particles may be formed throughout the polymerisation, a broad particle size distribution will result. If, however, no free emulsifier is available a monodisperse latex will result.

A particular use of this technique is the process whereby a 'seed' latex is first prepared from the same monomeric composition of the recipe using less than 25 per cent of the total monomers. Very low levels of emulsifier may be used with this technique with the remaining monomer and modifier added in increments at certain conversions. In the 'seed' process the incremental monomer becomes absorbed in the 'seed' particles and polymerisation takes place within the particle causing it to grow without further addition of emulsifier.

Continuous reaction

Continuous emulsion polymerisation is mainly employed commercially for the large scale production of synthetic rubber (SBR, nitrile and polychloroprene), although since 1970 it has become an increasingly important technique and is now also used for polybutadiene, polystyrene, carboxylic-SBR, ABS (acrylonitrile–butadiene–styrene) and PVC. The continuous system has advantages of productivity, process control and uniformity of product over the batch and semi-continuous systems.

The system was developed during World War II for the production of both styrene–butadiene (SBR) and acrylonitrile–butadiene (nitrile) rubber. A literature and patent review is available covering published work up to 1970,[4] and a later review deals in more detail with the theoretical aspects.[5]

Two systems may be used, the continuous stirred tank reactor (CSTR) system, consisting of 1–12 reactors, or a tubular system which may take the form of a continuous coiled tube or a series of unstirred tubes.

Reactors are similar to those described previously and are jacketed or contain internal coils for heating/cooling. They are agitated to maintain a homogeneous dispersion without too high a shear rate. The flow through the system is generally by feeding from the top of one reactor into the bottom of the next and so on. Reactants may be added to the first reactor or portions may be added between subsequent reactors. Alternatively, a pre-emulsified mixture may be added to the first reactor. The reaction product is taken from the last reactor and stripped to reduce residual monomer content to an acceptable level.

The continuous tube reactor system is one which, although not at present used to a large extent commercially, is gaining in popularity. The system may consist of a number of non-agitated, jacketed tubes which can be heated or cooled, or a continuous coiled tube in a heating/cooling medium, a holding vessel for pre-emulsified reactants and a means of emulsification which may be an homogeniser through which the resultant feed can circulate. The pre-emulsification vessel is agitated and can be controlled at the desired feed temperature. Initiator would normally be added just prior to entry to the reactor system to prevent possible polymer formation in the emulsification vessel. As with the CSTR system, other feed streams may be added along the reactor system.

A potential advantage of the tube system is that, if tube dimensions and flow rates are correctly determined, a system of the 'plug-flow' type is possible which combines the advantages of giving a product similar to that produced from a batch reaction with the productivity of a continuous system.

Post-reaction treatment
A factor common to all types of synthetic latex is that they must not contain high levels of residual monomers. Apart from the obvious health hazards, residual monomers can be responsible for odour problems in processing and end applications, and can have a deleterious effect on end-use properties. With the current awareness of health and safety requirements and the advent of more stringent legislation, specifications for residual monomer levels are constantly being reviewed. The maximum permissible level normally varies according to the type of monomer and/or the end-use application.

Most synthetic latices result from polymerisation reactions which are taken as far as possible to 100 per cent conversion, the main exception being high solids SBR latex used in spread and moulded foam applications. The reasons for lower conversions are dealt with in a later section. Where low conversions are employed, it is obviously economic to strip out and recover monomers for further use.

Latices taken to 'complete' conversion may contain up to 2 per cent by weight of residual monomers, and where very small quantities are involved it may not be economic to recover the monomer after stripping. This has resulted in methods being devised to reduce the residual monomer level to an absolute minimum during polymerisation, so that energy/cost requirements for stripping may be minimised or in some instances no stripping will be necessary. With most monomers the problem with removal of residual quantities is that the monomer is occluded in the polymer particles, which increases the difficulty of removal by stripping and of access by free radicals for further reaction.

Most latices normally require some post-reaction additives. This may only take the form of fixed or fugitive alkali addition for adjustment of pH, but could include biocides, antioxidants (particularly where double bonds are present in the polymer), reodorants, plasticisers or freeze-thaw additives.

USES

The largest industrial outlets for synthetic latices are surface coatings and adhesives, and many of the surface coating applications also depend on the adhesive properties of the polymer.

Other major outlets are paper including coating and impregnation; textile applications including carpet backing (secondary and pre-coat), sizing, semi-permanent finishing, stiffening and binders for non-woven fabrics; moulded foam and carpet spread foam; polishes, and bitumen and cement additives.

In certain carpet backing and paper coating applications, the actual compound or coating may contain as much as ten times by weight of filler or pigment as polymer, illustrating the extreme demands placed on the polymer.

End-uses of some latices will be referred to in this chapter and more specific applications are considered in later chapters.

MAJOR SYNTHETIC LATICES

The following sections of this chapter will be concerned with the major types of synthetic latex.

Carboxylic-SBR latices

During the past 20–30 years production of carboxylic-SBR latices has increased dramatically, with major consumption in textile and paper industries.

There are several different types of carboxylic latices available but the major ones are styrene–butadiene, styrene–acrylic, vinyl acetate–acrylate and butadiene–acrylonitrile. In Western Europe, 80 per cent of carboxylic latices are represented by carboxylic-SBR types and the remainder mainly by acrylic types.

The carboxy functional group confers important advantages such as the ability to use sulphurless cure systems, crosslinking with other functional monomers and high adhesive strength. The increased polarity of the polymer increases its compatibility with, and affinity for, polar substrates such as fibres or inorganic fillers.

The first recorded preparation of carboxylic latices was in the 1930s when a number of patents were published by a group of German workers. These described the preparation of butadiene–methacrylic acid copolymers, but it was not until the late 1940s that the first commercial product, a styrene–butadiene–acrylic acid terpolymer, was prepared and marketed in the USA.

Carboxylic-SBRs generally contain up to 5 per cent of carboxyl-containing monomer, and are available with 35 to 85 per cent styrene, giving a variation in degree of stiffness (handle).

Commercially, batch, semi-continuous or continuous processes are used with a preference for the semi-continuous method. Pressure vessels are required and reaction temperatures of 60–100°C are normally employed. Reaction is at an acid pH (2–4), using an anionic emulsifier (sodium salt of sulphonic acid or sulphosuccinic acid esters), taken to complete conversion using a thermally decomposible initiator (alkali persulphate). Temperatures are usually controlled to <100°C. This avoids high pressure, rapid depletion of radicals, and the formation of very highly gelled high molecular weight products because of the greatly reduced efficiency of chain transfer agents at high temperatures.

The acids used are α, β unsaturated mono- and dicarboxylic acids, the most common being acrylic, methacrylic and itaconic acids. The choice of acids depends on a number of factors which may include relative reactivity in copolymerisation, water solubility, end-use application of latex and, more importantly for commercial processes, cost effectiveness.

Functional groups other than carboxyl may be additionally introduced. These can be hydroxyl groups from hydroxyethyl esters of α, β ethylenically unsaturated acids such as 2-hydroxyethyl or hydroxypropyl acrylates or methacrylates, or amino groups from acrylamide or substituted acrylamide.

To obtain full benefit from the carboxyl groups, it is obviously necessary that the carboxylic acid constituent should be polymerised on the particle surface to form chemically bound carboxyl groups. Other possible modes of incorporation could be as follows:

(1) inside the particle, thereby burying the carboxyl group;
(2) in the aqueous phase to form water-soluble polymers that either adsorb onto the particle surface or remain in the aqueous phase;
(3) not to polymerise at all, but remain in the latex as monomer.

Therefore, to achieve optimum incorporation and disposition of carboxy groups, factors such as water solubility and the relative reactivities of the carboxylic acid components to copolymerisation have to be taken into account, and also the conversion to which the polymerisation is taken and whether or not a mixture of carboxy monomers is used.

Reactivity ratios of butadiene and the more commonly used carboxylic acids in copolymerisation with styrene are as follows:

Styrene (r_1)	Other monomer (r_2)	
0·78	Butadiene	1·39
0·23	Acrylic acid	0·25
0·20	Methacrylic acid	0·61
0·30	Itaconic acid	0·20

A number of variations in polymerisation procedure have been developed to ensure that the maximum amount of carboxyl-containing monomer is incorporated and that the carboxyl groups are concentrated at the surface of the polymer particles.

Commercially available carboxylic-SBR latices have the following properties:

Total solids content	50–55 per cent
Number average particle size	140–220 nm
Surface tension	>38 mN/m (dependent on end-use)
Viscosity	<300 mPa.s
Mooney (ML1+4)	80–160 (dependent on end-use)

As previously stated, the major outlets for carboxylic-SBR latices are in the textile (carpet), paper coating and adhesives industries. Within the carpet industry these latices may be used in pre-coat, secondary backing, backsizing, needlefelt and foam applications.

Vinyl acetate homo- and copolymer latices

Polyvinyl acetate (PVA) latices were developed commercially in Germany and USA in the 1930s, but growth in their use was slow until the 1940s. Since then growth has been steady, with an acceleration caused by the introduction of copolymers in the years subsequent to 1950.

Vinyl acetate differs from the majority of commercially used monomers in that it is appreciably soluble in water, having a solubility of 2·0–2·5 per cent by mass at ambient temperature, and in contrast to other water-soluble monomers, vinyl acetate monomer and polymer are completely miscible.

Industrially, PVA latices may be manufactured by batch, semi-continuous or continuous processes, with semi-continuous being the preferred method by a 'seeding' technique. Originally PVA latices were prepared using a protective colloid, polyvinyl alcohol, as the stabiliser. This has been retained to a large extent where levels up to 4 per cent by mass may be used for the preparation of latices with specific adhesive applications. Other latices utilise combinations of protective colloid, which may be polyvinyl alcohol or hydroxyethyl cellulose, with an anionic emulsifier of the sulphate or sulphonate type, and possibly a non-ionic emulsifier. A persulphate is normally used as the initiator, and polymerisation carried out at a pH of 4–6. A pH value outside this range can give polymer instability due to hydrolysis. Typically latices have solids contents up to 55 per cent and an average particle size, dependent on end-use, in the range 100–1000 nm.

Solid PVA has a second order transition point (T_g) of 25–28°C, making plasticisation necessary for some applications. Unlike vinyl chloride and its copolymers, this is readily achieved in emulsion with PVA and additions of >5 per cent by mass of dibutyl or dioctyl phthalate are necessary to give film forming properties and flexibility at ambient temperatures. However, plasticiser migration may occur under some conditions of use and a natural development was to substitute 'external' plasticisation by so-called 'internal' plasticisation whereby vinyl acetate is copolymerised with monomers to give added flexibility to polymer films. The type of comonomer used depends on availability, cost, ease of copolymerisation, and plasticising efficiency.

The main monomers used in copolymerisation are the vinyl esters of long chain fatty acids and trialkyl acetic acids (versatic acids), esters of acrylic, maleic and fumaric acids, and ethylene. In some applications, replacement of part of the vinyl acetate by vinyl chloride has given improvements in ethylene–vinyl acetate copolymers. Also, for certain applications, small quantities of unsaturated carboxylic acid monomer (c.f. carboxylic-SBR latices) may be included, and self-crosslinking types may incorporate small quantities of other functional monomers.

Reactivity ratios of some of the more common comonomers with vinyl acetate are as follows:

Vinyl acetate (r_1)	Other monomers (r_2)	
0·01	Diethyl fumarate	0·44
0·17	Diethyl maleate	0·04
0·29	Methyl acrylate	6·70
0·01	2-Ethylhexyl acrylate	6·70
0·04	n-Butyl acrylate	5·50
1·08	Ethylene	1·07

Copolymerisation of vinyl esters proceeds readily to give homogeneous copolymers. Copolymerisation with acrylate esters, however, requires to be carefully controlled since the acrylate esters are more reactive than vinyl acetate. Use of the 'seed' technique produces reasonably homogeneous copolymers. Copolymer contents vary according to the effectiveness of the copolymer, and may vary from 10 per cent for ethylene up to 50 per cent for vinyl versatate. On a mass basis the plasticising effect of ethylene is greater than that of other comonomers.

Similar polymerisation systems are employed to those used in the preparation of PVA latices, with surfactant systems varying according to end-use requirements. However, for the preparation of ethylene copolymers more expensive pressure reactors are required.

Vinyl acetate copolymers are available at solids contents up to 55 per cent and average particle size up to 1500 nm. The higher solids and larger particle size latices are normally used in adhesive applications. Latices for paint, textile and paper coating applications generally have solids contents up to 50 per cent, with an average particle size of 100–200 nm.

The main outlets for vinyl acetate homo- and copolymers are paints, adhesives, non-wovens, paper coating and carpets, with the market in Western Europe accounting for 49 per cent in paints and 44 per cent in adhesives.

In many applications, vinyl acetate polymers and copolymers are preferred to those containing butadiene as they exhibit superior resistance to oxidation and UV light. Their stability, however, may fall on storage due to a decrease in pH arising from hydrolysis of residual vinyl acetate monomer.

Acrylic latices

Acrylic latices are those latices containing polymers whose major components are derived from esters of acrylic or methacrylic acid, and include styrene/acrylates.

Acrylic ester polymer latices have been commercially available since 1925–1930 when they were used as base coats in the finishing of leather. Today large volumes of a variety of copolymer and multipolymer latices are used in a wide number of applications.

The types of acrylic latices are dependent on the required applicational properties, the main properties being flexibility, hardness, flow, and temperature of film formation, all of which are all related to the second order transition point (T_g).

The most widely used monomers for latices are ethyl acrylate, methyl methacrylate and *n*-butyl acrylate. Other monomers used include ethyl methacrylate, 2-ethylhexyl acrylate, 2-ethylhexyl methacrylate, *n*-butyl methacrylate, methyl acrylate, isobutyl acrylate and isobutyl methacrylate.

Table 3.2 gives the glass transition temperatures and monomer reactivity ratios with styrene (M_2) of some of the more common acrylic monomers (M_1).

TABLE 3.2
Glass transition temperatures and monomer reactivity ratios with styrene

M_1	r_1	r_2	$T_g(°C)$
Methyl acrylate	0·14	0·68	8
Ethyl acrylate	0·16	1·01	−22
n-Butyl acrylate	0·21	0·80	−54
2-Ethylhexyl acrylate	0·26	0·94	−85
Methyl methacrylate	0·46	0·52	105
Ethyl methacrylate	0·41	0·53	65
n-Butyl methacrylate	0·47	0·52	20
Isobutyl methacrylate	0·40	0·55	48

Functional monomers may be copolymerised to introduce reactive pendant groups for altering solubility characteristics or for crosslinking with a catalyst at ambient temperature or by heating. Typical reactive sites are hydroxyl groups introduced by copolymerisation of 2-hydroxyethyl or propyl acrylates or methacrylates. Other reactive groups include amide, methylol- or alkoxymethylolamide, carboxyl and epoxy.

Minor proportions of styrene, vinyl esters or other non-acrylic monomers may be included to further modify hardness and flexibility properties. Also minor proportions (1–2 per cent) of polymerisable acids such as acrylic, methacrylic or itaconic acid or the half-ester of a dibasic acid may be incorporated to act as internal emulsifiers and improve freeze–thaw stability properties. Acid groupings also aid pigment wetting. Excess acid groups are sometimes used and this leads to stable emulsions of low pH. These products are alkali soluble and are used in high gloss emulsion paints.

The various monomer combinations used give specific characteristics to polymer film properties. Polyacrylates generally have high elongation and low strength, but polymethacrylates give high strength and low elongation. Film flexibility is given by ethyl acrylate, butyl acrylate and 2-ethylhexyl acrylate, and hardness by methyl methacrylate, methacrylic acid, acrylic acid and styrene.

Polymerisation is usually carried out by the semi-continuous technique. Emulsifiers commonly used are anionic, i.e. alkyl sulphates, alkylaryl sulphonates and phosphates, and combinations of these with nonionics such as polyethylene oxide condensates with long chain alcohols. The particular combinations of emulsifiers are governed by factors such as particle size requirements and long-term storage and viscosity stability. Initiators are of the water-soluble type, such as ammonium or potassium persulphate, and may be used in a redox system with sodium bisulphite or thiosulphate to give lower reaction temperatures. As with other synthetic latices, molecular weight and physical properties depend on the combination of monomers, and reaction conditions such as rate of monomer addition and reaction temperature, emulsifying system and the concentration used.

Acrylic latices are available with solids contents in the range 40–55 per cent, and particle size 100–500 nm. They have been used in emulsion paints for over 20 years and have good adhesion under adverse conditions, retention of colour and gloss, flexibility and alkali resistance. Paint represents the major outlet, accounting for about 50 per cent of Western European output. Other areas of use include adhesives for high speed packaging, fabric back-coating and finishing, and for bonding non-wovens, paper and board (beater addition), surface sizes, paper coating and saturants, water-thinnable inks, and cement additives.

Acrylics have certain advantages over vinyl acetate copolymers in paints in that they are generally of smaller particle size and have better pigment binding characteristics and flow properties, but they may be inferior in emulsion levelling.

Styrene, butadiene and copolymer latices

This class represents a range of latices which give products varying from the soft and rubbery polybutadiene to the hard and resinous polystyrene, with intermediate products having styrene levels of 23 to 85 per cent.

Polybutadiene latices find use mainly as the base latices for the preparation of ABS plastics, styrene and acrylonitrile being grafted onto the rubbery backbone. For this application the polybutadiene must have very specific properties, i.e. particle size and distribution, gel content and swelling index. Reactions are generally batchwise of long duration (up to 60–70 h), using rosin acid soap and persulphate initiator at 70–90°C for 80 per cent conversion.

Polystyrene latices are normally used as components of other latex systems to give stiffening or reinforcing properties. If used alone, formulations generally include a plasticiser. Latices may be prepared by batch or continuous methods. Soap systems include sulphates, sulphonates, fatty and rosin acid salts with persulphate as the initiator. Reaction temperatures are normally in the range 70–100°C, and the conversion is 100 per cent. Particle sizes are normally in the range of 100–150 nm, but for polish applications latices of very small particle size (<50 nm) are used.

The so-called 'high styrene resins' are styrene/butadiene latices with a styrene content of 80–85 per cent. They are produced using recipes and conditions similar to those used for polystyrene, with rosin acid and fatty acid salts being the preferred emulsifiers. As for polystyrene latices these materials find wide application as stiffening and reinforcing agents, and because of their lower softening points and increased thermoplasticity they also find application as impregnants for use in thermoforming processes.

The largest volume production of SBR latices gives material of 23–25 per cent bound styrene, with the bulk being further processed to produce the many types of dry rubber. Considerable amounts, however, of these SBR latices find use in foam rubber and adhesive applications. The processes for these products were developed during World War II for the manufacture of synthetic rubber, and the recipes and procedures are still being used. Latices are prepared by a continuous process with 10–12 reactors. Processes are termed 'cold' (5°C) or 'hot' (50°C) with the 'cold' process accounting for the larger volume, and also for the material used as a latex. Emulsifier systems may be fatty acid or disproportionated rosin acid soap or a mixture of both, with a minor proportion of a formaldehyde–naphthalene sulphonate condensate. Redox systems are used as free radical sources, with a hydroperoxide as the oxidising component and sodium formaldehyde sulphoxylate (SFS) as the reducing component, together with ferrous sulphate and a chelating agent (ethylene diamine tetra-acetic acid, EDTA). Tertiary dodecyl mercaptan is used as a chain transfer agent.

Reactions are taken to a predetermined monomer conversion by the addition of a shortstop agent, generally sodium dimethyl dithiocarbamate. Conversion is usually 60–65 per cent for the 'cold' process, giving a linear polymer with limited branching and crosslinking. Higher conversions (65–72 per cent) such as those used in 'hot' reactions give a polymer which may contain some branching and crosslinking. Unreacted butadiene is removed by flash distillation, and unreacted styrene by steam distillation. These two monomers are recovered and recycled.

Latex prepared by this method normally has a solids content of 30–35 per cent and a particle size of 65–75 nm, with a monodisperse distribution. Use of this material is limited mainly due to solids content which can be increased by evaporation, but at solids values in excess of 40 per cent, the viscosity becomes too high. The latex, however, has applications in some adhesive areas.

For spread and moulded foam applications of SBR latices, solids contents must be greater than 64 per cent with viscosity not exceeding 1500 mPa.s. To obtain a latex of this type (high solids latex, HSL) it is necessary to modify the particle size and distribution. Although specially developed recipes have been used commercially for the production of large particle size SBR latices, polymerisation rates are very slow, and generally, conventional recipes are preferred with the particle size modified by post-treatment. With the separate processes of polymerisation, increase in particle size (agglomeration), and evaporation, the various parts are relatively independent of each other and this enables the recipe and subsequent stages to be adjusted to give the most economic conditions and optimum properties of the final latex.

There are a number of methods available for the agglomeration of latices to produce the required particle size and distribution to allow the latex to be concentrated to a high solids content. These are solvent addition, soap neutralisation, electrolyte addition, freeze agglomeration, chemical agglomeration and pressure agglomeration. Only the latter three are used commercially. In freeze agglomeration, the latex is frozen and then thawed — agglomeration is accompanied by a high coagulum level. For chemical agglomeration, polymers such as polyvinyl methyl ether, polyethylene oxide or polyethylene glycol must be added — these remain in the concentrated latex. Pressure agglomeration is a cheap, versatile, easily controllable process with little or no coagulum formation, and is the most widely used commercial method. The process is continuous and requires only relatively simple equipment, agglomeration being achieved by passage through an homogenising valve, contrary to the normally accepted use of an homogeniser which is for reduction of particle size.

Various parameters are known to affect the extent of agglomeration and the facility with which it can be carried out. These are pH, solids content, temperature, agglomeration pressure and soap-to-rubber ratio. Increased agglomeration is obtained by decreasing the pH, increasing solids, decreasing temperature, increasing homogenisation pressure, decreasing soap-to-rubber ratio or a combination of these.

Commercially, the base latex is concentrated to a solids content of 38–40 per cent by evaporation to increase throughput of the homogeniser and also to allow other conditions (pH, pressure, etc.) to be less severe. Concentrated latex at 32–35°C is pH adjusted (if necessary) by the addition of a dispersion of sodium silicofluoride (SSF) and pumped through the homogeniser (pressure 3·5–3·9 MN), and then evaporated to the final solids level of 64–68 per cent with a viscosity of 800–1200 mPa.s, a number average particle size of 100–160 nm and a weight average particle size of 300–400 nm.

Foams made from this type of latex have good elongation and tensile strength, high hardness and low compression set. For most end-uses, however, improved hardness is required with retention or improvement of other properties. This is achieved by incorporation of a reinforcing latex such as polystyrene or a high styrene resin, which may be incorporated by diagglomeration or blending.

High solids SBR latices are used in the manufacture of carpet spread foam of the non-gel, ammonium acetate gel and SSF gel type, and moulded foam from the SSF gel and Talalay processes.

Finally in this section should be included an important terpolymer of styrene and butadiene with 2-vinyl pyridine with incorporated levels of 15, 70 and 15 per cent, respectively. These are speciality materials used as tyre cord adhesives and in the belting industry. Preparation is by a batch method using programmed temperature control and incremental soap addition; the agitation pattern of the reactor is a very important factor. Vinyl pyridine latices have to be made to a demanding specification which includes total solids, pH, particle size, surface tension, mechanical stability and Mooney viscosity. The soap system comprises rosin acid salt and sodium salt of a naphthalene sulphonic acid–formaldehyde condensate, with a persulphate as the free radical initiator. The reaction temperature is generally in the range 40–70°C with a reaction time of 30–40 h when taken to full conversion.

Nitrile latices

The name nitrile applies to the class of materials which are copolymers of unsaturated nitriles with dienes. The most common are copolymers of acrylonitrile and butadiene and these are described in this section.

Nitrile rubber was prepared using an emulsion polymerisation process as early as 1930, and was commercially available as Buna N in the late 1930s in Germany. This was followed by production in the USA around 1940.

The polymerisation process is analogous to that used for SBR latices; 'hot' and 'cold' recipes may be used. Emulsifying systems, as for SBR latices, are predominantly salts of fatty acids or disproportionated rosin acids, or mixtures of these with minor amounts of other anionic or non-ionic soaps. Modifiers are usually tertiary or normal dodecyl mercaptans, with initiation by redox systems, allowing lower temperatures and more controlled reactions. Gel levels may be kept to a minimum by control of temperature and conversion level, although the temperature is a far less important variable than with SBR latices.

The reactivity ratios for free radical copolymerisation of acrylonitrile (M_1) and butadiene (M_2) are given as $r_1 = 0.03$ and $r_2 = 0.30$, meaning that normally the polymer composition changes with conversion. However, a charge of butadiene/acrylonitrile of mole ratio 1·67 (azeotropic composition) will give the same copolymer composition as the charge composition.

Acrylonitrile levels may be in the range 10–45 per cent with an average value of about 33 per cent. Nitriles are commonly referred to as having a low, medium or high acrylonitrile content, indicating levels of 25, 33 or 45 per cent, respectively. The monomer ratio is probably the most important variable, with the polarity of the copolymer increasing with increasing acrylonitrile content, affecting oil resistance, flexibility, and adhesion to polar substrates.

A large proportion of nitrile latices produced are of the modified type, which contain other reactive groupings (possibly up to 10 per cent), and which may give self-crosslinking properties or allow the polymer to crosslink with other polymers or additives. The main groups used are carboxyl, amide and substituted amide. These types may be prepared in a manner analogous to that described for carboxylated SBR latices.

Apart from the variation in acrylonitrile content, nitrile latices are available at solids contents up to 60 per cent (average range 40–50 per cent), particle sizes 40–200 nm, Mooney viscosity (ML_{1+4}) 50–150.

Nitrile latices are used in applicational areas where oil and abrasion resistance and high binding power to polar substrates are required. The main applications are in the textile (non-wovens) and paper industries, glove dipping, surface coatings and adhesives. There are certain advantages are the relatively slow drying and weaker bond strengths when application and absence of fire and toxicity hazards. The main disadvantages are the relatively slow drying and weaker bond strengths when compared to solvent-based products.

Polyvinylidene chloride latices
These latices have been known since 1930–1940, when the original work on emulsion polymerisation was carried out at Dow Chemical Co., USA.

Vinylidene chloride (VDC) monomer is different from those considered so far in that the polymer shows only limited miscibility with its own monomer and this can lead to complications in polymerisation. Also polyvinylidene chloride (PVDC) can exist in both amorphous and crystalline forms, and the latex polymer itself may change from one form to the other.

The homopolymer of VDC has little commercial value, the most important latices being copolymers with vinyl chloride, acrylonitrile, and various alkyl acrylates and methacrylates. Reactivity ratios of VDC (M_1) in copolymerisation with some comonomers (M_2) are as follows:

Monomer, M_2	r_1	r_2
Methyl acrylate	1·00	1·00
Methyl methacrylate	0·24	2·53
Vinyl chloride	3·20	0·30
Acrylonitrile	0·37	0·91
Styrene	0·14	2·00
Vinyl acetate	6·00	0·10

Proportions of comonomer present may range from about 50 per cent down to 7 per cent.

Polymerisation is by the semi-continuous method, with glass-lined or high grade stainless steel pressure reactors being necessary because of the possibility of hydrochloric acid generation. The surfactants used are generally anionic or mixtures of anionic and non-ionic types. These may be the sodium salts of alkyl aryl sulphonates, alkyl esters of sulphosuccinic acid or fatty alcohol sulphates for the anionic type and ethoxylated alkyl phenol for the non-ionic type.

Initiator systems are normally the redox type, such as ammonium persulphate/sodium *meta*bisulphite, with reaction temperatures of 40–50°C to give latices of solids content up to 55 per cent and particle size of 100–250 nm.

A large range of properties is obtainable from PVDC latices and can be varied by the usual techniques of copolymer composition and mode of manufacture. These latices are, however, susceptible to colour formation which is believed to be caused by dehydrochlorination leading to conjugated unsaturation.

As mentioned previously, PVDC copolymers can exist in both amorphous and crystalline forms. This property is exploited in barrier coating applications where the polymer can be designed so that it is amorphous in the wet state with good film forming properties, but can then revert to a crystalline form on or after application giving improved barrier, non-blocking, and slip properties.

The main uses of PVDC latices are in barrier coatings for paper, film, aluminium foil, board, etc., in textile and carpet applications because of flame retardant properties, and in paints, particularly fire retardant and vinyl silk types.

Barrier and other properties of PVDC polymers are related to the VDC content of the polymer. High VDC (90 per cent) content polymers are used where a high barrier against gas and water is required. For board coating, where high flexibility and improved light fastness is necessary, polymers of 80 per cent VDC with a soft comonomer are used. Intermediate VDC (50 per cent) levels have inferior barrier properties, but show improved toughness and water resistance, and less blocking in the film compared with acrylic copolymers.

Polyvinyl chloride latices

Vinyl chloride monomer (VCM) is similar to VDC in that polyvinyl chloride (PVC) is insoluble in the monomer.

Emulsion polymerisation of VCM was developed and used in Germany during World War II. Polymerisation was carried out in closed rotating autoclaves at 40–50°C for 24 h. The resultant latex was spray dried and the product used as a resin. Subsequent improvements led to a continuous process.

Polyvinyl chloride is the most common thermoplastic material and is used in many varied applications due to its strength and abrasion resistance, resistance to flammability, chemical and water resistance, and its ability to be softened by plasticisers to give a wide range of hardnesses. Commercially today, two types of PVC homopolymer are produced. One is used in general latex applications, and the other for 'plastisol' production, which is beyond the scope of this book.

Polyvinyl chloride homopolymer is hard and non-film forming, and when plasticised often requires temperatures in the region of 150°C for complete fusion. However, fusion temperatures may be reduced by copolymerisation with other monomers such as acrylates, maleates, fumarates, or ethylene. Flexibility and softness are increased with little loss in toughness and abrasion resistance. When external plasticisers are used these are normally higher phthalates or phosphate esters.

Reactivity ratios for some copolymer systems with vinyl chloride are as follows:

Monomer, M_2	r_1	r_2
Butyl acrylate	0·07	4·40
Ethylene	1·85	0·20

Methyl acrylate	0·12	4·40
Methyl methacrylate	0·02	15·00
Vinyl acetate	1·68	0·23
Vinylidene chloride	0·30	3·20
Diethyl maleate	0·77	0·01
Diethyl fumarate	0·12	0·47
Dibutyl maleate	1·40	0·00

Preparation of PVC latices is by the 'seed' process and semi-continuous method in a batch reactor. Recipes and procedures are very similar to those used for PVDC. Emulsifiers are generally alkylaryl sulphonates, sodium lauryl sulphate, or salts of substituted sulphosuccinic acids, with initiators generally of the water-soluble redox type, i.e. persulphate/bisulphite. As with PVDC, lower temperatures are preferred to keep operating pressures down and obtain higher molecular weight polymers. Operating temperatures are generally 50–60°C with reaction times of up to 20 h.

Latices are available with solids contents up to 58 per cent and particle sizes in the range 80–200 nm. Homo- and copolymer latices may be obtained in the plasticised and unplasticised form and in some instances sufficient copolymer is present to give internal plasticisation.

Polyvinyl chloride latices find applications in textiles, paper and board coating, and beater addition and impregnation, particularly where heat sealability properties may be exploited. Use is also made of the fire retardancy characteristic in many applications.

In recent years, copolymers with ethylene have become available, which also may contain a minor proportion of a third monomer with carboxyl or amide groups. These products are claimed to be self-extinguishing and will complement the effects of fire-retardant additives. They are compatible with PVDC, acrylic and vinyl acetate latices, and find use in carpet backing, barrier coating, beater addition, saturants, industrial coatings, adhesives, and water-based inks.

Polychloroprene latices

Polychloroprene (CR) is the polymer of 2-chlorobutadiene, and in 1931 was the first synthetic rubber to be widely available commercially, and by 1934 was also available for use in the latex form. There are today about 15–20 different types of chloroprene homo- and copolymer latices available.

Generally CR latices are prepared batchwise using an anionic surfactant system of salts of rosin or disproportionated rosin acids, with potassium persulphate as the initiator at 40–50°C. Conversion is monitored by the specific gravity change, and the polymerisation may be short-stopped or allowed to go to full conversion depending on the type of latex.

Reactivity ratios with a number of comonomers are given below:

Monomer, M_2	r_1	r_2
Acrylonitrile	6·93	0·03
Styrene	5·20	0·00
Butadiene	2·90	0·00
Methacrylic acid	2·68	0·15
Vinyl acetate	50·00	0·10
Methyl acrylate	11·10	0·08

The polymers in the majority of CR latices are of the 'gel' type, and are partially crosslinked, insoluble in aromatic solvents and have very high Mooney viscosity. There are a small number of latices available which contain 'sol' type polymer, characterised as soluble in aromatic solvents and having a very low Mooney viscosity. These types of latex are produced by making a sulphur-modified polymer by incorporation of sulphur in the polymerisation recipe leading to polysulphide linkages. The polymerisation is short-stopped at about 90 per cent conversion by the addition of a thiuram disulphide. During an ageing period at ambient temperature the sulphide linkages are cleaved by the thiuram disulphide which thereby reduces the molecular weight.

Polychloroprene latices generally have solids contents in the range 35–60 per cent, with a high pH (12·0), and particle size in the range 50–190 nm. Different types of latex are produced by variation of polymerisation parameters such as temperature, emulsifier, initiator and modifier, leading to 'sol' and 'gel' types and varying degrees of crystallinity.

Latices may be classified as general purpose or speciality. General purpose grades are homopolymers, anionic and of the 'gel' type. Speciality grades are both 'gel' and 'sol' types, and also include copolymers with acrylonitrile, styrene and methacrylic acid. A cationically stabilised latex is also available with a quaternary ammonium salt as the stabiliser. Higher solids latices (up to 60 per cent) of both classes are obtained by alginate creaming of the respective standard solids latices.

Polychloroprene products are generally similar to natural rubber, but superior in resistance to oils, solvents, ozone, sunlight, oxidation and flex-cracking. They find application in a wide variety of end-uses including bonded fibres, coatings, adhesives, treated paper, concrete additives, dipped goods, sealants, foam and modified bitumens.

Artificial latices

Artificial latices are those latices formed from polymers produced by processes other than emulsion polymerisation and, therefore, do not strictly belong to the class of synthetic latices. Their main method of preparation is by the solution emulsification technique. In this process the polymer is dissolved or swollen in a solvent, emulsified in water, and the solvent removed by distillation. The solvent used for polymer dissolution or the solvent/water azeotrope must have a boiling point below that of water, and for economic and environmental reasons the solvent should be readily separable from water. The crude emulsion is subjected to high shear to give an emulsion with particles of less than 1000 nm. The choice of emulsifier is an important part of the operation, as a stable emulsion must be formed initially that will withstand the solvent stripping operation and ultimately give a mechanically stable latex. There are two main types of artificial latex available which are produced by this method, butyl and *cis*-1,4-polyisoprene latices.

Butyl rubber is a copolymer of isobutylene with a minor proportion of isoprene. Butyl latex prepared from this rubber has a solids content greater than 60 per cent, an average particle size of 500 nm (with an anionic emulsifier), and possesses good mechanical, chemical and freeze-thaw stability. This type of latex finds application in non-woven binders, fabric finishing and coating, protective coatings, adhesives, sealants, and paper saturants.

Stereoregular *cis*-1,4-polyisoprene is produced by a solution polymerisation process. The latex is prepared by the solution emulsification process and is available at greater than 60 per cent solids content with an average particle size of 700 nm at pH 10, and has a high mechanical, but low chemical stability. The latex shows two major differences from natural rubber latex in that the polymer is mainly linear and free of branched material and micro-gel, and the non-rubber constituents are only present in very low quantities. The latex may be used in such applications as foam, dipped goods and adhesives.

Polyurethane (PU) latices are becoming more readily available. These cannot be prepared by emulsion polymerisation as polyurethanes are normally prepared under strictly anhydrous conditions. However, methods have been developed for the preparation of non-ionic, anionic and cationic latices.

Non-ionic PU latices may be produced from water-soluble polyethylene glycol ethers and diisocyanates. Anionic latices contain acid groupings, and cationic latices positively charged groups in the polymer. By careful selection of the PU composition, stabilising/wetting agents and the mode of preparation it is possible to produce a range of PU latices. Solids contents are generally up to 50–60 per cent, with particle sizes in the range 50–500 nm. In general, applications utilise the characteristic high tensile and tear strength, abrasion resistance and solvent resistance of PU, and offer a wide range of hardness, flexibility and elastic properties, which can be attained by varying molecular structure without the use of external additives. These latices find application in dipped goods, surface finishes, soil release properties (textiles, carpets), binders and finishes for non-wovens, improvement of shrinkage, and crease resistance in permanent.press materials.

FUTURE DEVELOPMENTS

Since the commercialisation of polyvinyl acetate latices in the 1950s, the synthetic latex industry has enjoyed a period of substantial growth. Today, about 90 per cent of the market is accounted for by PVA and copolymers, SBR, and acrylics. Of the many industries using these latices, more than 85 per cent are used in what may be termed non-durable products, such as non-wovens, carpets, surface coatings, adhesives, paper and textiles. During the past 5–10 years there has been a slowing down of the growth rate and it has been predicted in the USA that the growth rate over the next 20 years will be fairly close to that of the Gross National Product.

Trends in the preferred types of synthetic latices are not easy to predict and in most instances are closely related to cost/effectiveness, with the bulk products, SBR, VA copolymers and acrylics being interchangeable for a large number of applications. Cost is governed to a certain extent by scale and mode of manufacture, but the most important factor is raw material (monomer) cost.

It is thought that future developments will see more bulk production of multipurpose latices with a reduction of the large number of low tonnage speciality products, and a move towards continuous processes, offering economy and consistency of production, with a far greater emphasis on energy conservation.

In view of current and future legislation on health and safety, the trend of replacing solvent-based by water-based products (c.f. adhesives) will continue with efforts directed towards obtaining products of at least equivalent performance.

Fire retardancy, flame resistance and smoke emission are properties which have received an increasing amount of publicity in recent years, and there is much research and development work in progress in these areas. This will undoubtedly lead to polymer modifications or new monomer combinations to meet the stringent legislative requirements.

REFERENCES

1. DUCK, E. W., *Encyclopaedia of Science and Technology*, Vol. 5, Interscience, New York, 1966, 801.
2. VANDERHOFF, J. W., *Vinyl Polymerisation Pt. II*, Chapt. 1, Marcel Dekker, New York, 1969.
3. BLACKLEY, D. C., *Emulsion Polymerisation*, Applied Science Publishers, London, 1975.
4. FEAST, A. A. J., *Rep. Prog. Appl. Chem.*, 1971, **56**, 45.
5. POEHLEIN, G. W. and DOUGHERTY, D. J., *Rubber Chem. Technol.*, 1977, **50**, 601.

Chapter 4

LATEX SPECIFICATIONS AND TEST METHODS

K. O. CALVERT
Dunlop Ltd, Birmingham, UK

This chapter describes latex specifications, codification of synthetic rubber latices and test methods for all types of latex. Reference is made, where possible, to documents published by the International Standards Organization (ISO). In other cases, either a national standard is cited or attention is drawn to the relevant technical literature.

Standards published by the International Standards Organization are available in English, French and Russian. They all have a unique 'ISO' number, e.g. ISO 2004, and are available from national standards bodies. Since several of the standards have been revised, the reader should ensure that the latest edition is obtained.

Test methods are considered under the broad headings of sampling and coagulum content, concentration, alkalinity and pH, mechanical stability, natural latex quality, chemical stability, polymer composition, volatile unsaturates, particle properties, chemical tests and other methods.

Throughout this chapter emphasis is placed, not on detail, but on salient aspects. For full details, the appropriate reference should be consulted.

SPECIFICATIONS FOR NATURAL RUBBER LATEX

Specifications for centrifuged or creamed, ammonia-preserved, natural rubber (NR) latices are set out in ISO 2004. Those for evaporated NR latices are the subject of ISO 2027. At the time of writing (1980), both of these documents were in second editions.

ISO 2004 deals with three types of centrifuged latex (HA, LA and XA), and two types of creamed latex (HA and LA). ISO 2027 covers three types of evaporated latex, two of which are preserved with potassium hydroxide. Reference was made in Chapter 2 to all these variants of NR latex.

Since type LA (centrifuged) is of increasing commercial importance, and because it embraces at least five sub-types (see Chapter 2), it is appropriate to concentrate on its specification. The more important requirements for type LA latex are as follows:

Minimum dry rubber content	60·0 per cent
Maximum alkalinity (as ammonia)	0·29 per cent
Minimum mechanical stability	650 s
Volatile fatty acid (VFA) number	As agreed, not exceeding 0·20
Potassium hydroxide number	As agreed, not exceeding 1·0

Total solids content is an optional requirement. Other characteristics that are specified are non-rubber solids, coagulum content, copper and manganese contents, sludge content, colour on visual inspection and odour after neutralisation with boric acid. It is unusual for any of these other characteristics to cause concern.

The apparently strange maximum alkalinity of 0·29 per cent for LA latex arises simply because the minimum alkalinity for XA latex is 0·30 per cent. In practice, the alkalinity of commercial LA latex is usually within 0·02 of 0·20 per cent.

Mechanical stability is probably the characteristic of most interest to the latex consumer and provision is made in the specification for a greater minimum mechanical stability than 650 s to be agreed. Occasionally, a maximum mechanical stability is also negotiated. Before 1979, a lower minimum mechanical stability was specified (540 s).

Another provision in the specification is to allow the potassium hydroxide number to exceed the specified value in the case of latex containing boric acid. This excess is, however, limited to an amount equivalent to the boric acid content of the latex. It is also intended to add a requirement for zinc stability to the specification but, despite considerable study over many years, a suitable test method has still to be devised.

The specifications for types HA and XA are identical to that of LA latex except as regards alkalinity. The requirements for creamed latex are also the same with the exception of minimum dry rubber content (64·0 per cent) and alkalinity.

The specifications for evaporated NR latices require a minimum total solids content (instead of dry rubber content), allow higher non-rubber solids and sludge content maxima, and do not include the potassium hydroxide number. Alkalinity is expressed as potassium hydroxide or ammonia depending on the type of evaporated latex.

CODIFICATION OF SYNTHETIC RUBBER LATICES

The international codification scheme for synthetic rubber latices is based on ISO 1629. This nomenclature document specifies the correct letters to represent the chemical family of the rubber in the latex. For example, SBR denotes styrene–butadiene rubbers, PSBR represents vinylpyridine–styrene–butadiene rubbers and XSBR describes carboxylic–styrene–butadiene rubbers.

In the codification of synthetic rubber latices (ISO 2438), the nomenclature is extended by the addition of two digits and, where relevant, a suffix letter. The first digit, which may be from 1 up to 7, represents the nominal total solids content of the latex. The second digit, which may be from 0 up to 6, represents the nominal bound (polymerised) comonomer content.

In the case of a styrene–butadiene rubber latex that is reinforced with polystyrene or a copolymer of butadiene and styrene, the bound comonomer content is considered to include the bound styrene content of the reinforcing polymer. In this situation the suffix letter Y is added to denote that the latex is reinforced.

To illustrate application of the international codification scheme, Table 4.1 shows four examples that refer to latices which are produced in substantial volume. Specifications for these types of synthetic rubber latex are considered in the next section.

TABLE 4.1
Examples to illustrate the application of the international codification scheme

Polymer in latex	Nominal total solids content (per cent)	Nominal bound styrene content (per cent)	Code
Styrene–butadiene rubber	68	25	SBR 62
Polystyrene-reinforced styrene–butadiene rubber	65	35	SBR63Y
Vinylpyridine–styrene–butadiene rubber	42	15	PSBR41
Carboxylic–styrene–butadiene rubber	55	45	XSBR54

For a full appreciation of the codification scheme for synthetic rubber latices, the reader is referred to ISO 2438, the second edition of which is due for publication in 1981.

SPECIFICATIONS FOR SYNTHETIC RUBBER LATEX

There are no international specifications for synthetic rubber latices. However, there are British Standard specifications (BS 4661) for five types of large-volume, styrene–butadiene latex. This standard specifies the tolerance allowed on the nominal values stated for total solids content, (total) bound styrene content, viscosity and pH of the latex. Also specified are limiting requirements for coagulum content, volatile unsaturates and non-polymer solids (i.e. the difference between total solids content and dry polymer content).

For SBR63Y latex, BS 4661 permits tolerances of,

± 1.5 per cent total solids,
± 2.0 per cent total bound styrene,
± 40.0 per cent on viscosity at minimum total solids, and
± 0.5 pH.

Thus, for example, if the supplier states the nominal total solids content as 65.0 per cent, the actual total solids content allowed is at least 63.5 per cent and not greater than 66.5 per cent. Similarly, if the nominal viscosity at minimum total solids (63.5 per cent) is declared as 500 mPa.s, the actual viscosity at this concentration is required to be at least 300 mPa.s and not more than 700 mPa.s.

The following limiting requirements for SBR63Y latex are also specified in BS 4661:

Maximum coagulum content	0.10 per cent
Maximum volatile unsaturates (calculated as styrene)	0.15 per cent
Maximum non-polymer solids	6.5 per cent of total solids

In addition, BS 4661 includes specifications for SBR62 'cold' latex, SBR42, PSBR41 and XSBR54 latices. Their permitted tolerances and limiting requirements are similar to those for SBR63Y latex.

SAMPLING AND COAGULUM CONTENT

Sampling
Sampling of rubber latex is the subject of ISO 123. This standard describes the procedures to be adopted to obtain a representative sample of latex contained in drums or in bulk (tanks and tank cars). It is emphasised that the material of the container in which the latex sample is kept must be impermeable and chemically resistant to the latex. Space needs to be left in the sample container to allow for thermal expansion of the latex and the container should be kept closed whenever possible. In handling latex samples, any operation that introduces air, such as cascading the latex, is to be avoided since foam on the latex surface skins rapidly.

Coagulum content
ISO 123 specifies that if the coagulum content of the latex exceeds 0·05 per cent, it shall be filtered before it is subjected to other tests. The coagulum content of rubber latex is defined as the material retained on stainless steel wire cloth with an average aperture width of 180 ± 15 μm under the conditions of the test (ISO 706). In this test, 200 g of latex is diluted with soap solution and passed through the specified filter. After thorough washing, the residue on the filter is dried to constant mass at 100°C.

It should be noted that the coagulum content of natural rubber latex is now expressed as a percentage by mass of the latex, in the same manner as synthetic rubber latex. Prior to 1976, it was calculated on the total solids content of the latex

Gross particle content
In the case of aqueous dispersions of plastics polymers and copolymers, the gross particle content is determined in similar manner to the coagulum content of rubber latex. However, differences are that gauzes of five mesh sizes are allowed (the largest of which is 180 μm), dilution and washing are carried out with water and the residue is dried at 105°C (ISO 4576).

TEST METHODS FOR CONCENTRATION

The concentration of latex is usually expressed as its total solids content. However, the dry rubber content of natural rubber latex is an important property and the dry polymer content of certain types of synthetic rubber latex is a useful criterion.

Total solids content

The total solids content of rubber latex is determined by drying approximately 2·0 g of latex to constant mass (ISO 124). The drying temperature is 100°C or 70°C for natural rubber latex and the same options or 125°C at reduced pressure (less than 20 kPa) for synthetic rubber latex.

For aqueous dispersions of plastics polymers and copolymers, a similar method is used for determination of the 'residue at 105°C' (ISO 1625). However, less latex is tested, it is weighed more accurately and there is no confirmation that the residue is dried to constant mass.

Dry rubber content

The dry rubber content of natural rubber latex is determined by coagulating approximately 10 g of latex, after dilution with water, with dilute acetic acid. The coagulated rubber is digested, thoroughly soaked in water and pressed to a thickness of 2 mm or less before it is dried to constant mass at 70°C. The second edition of ISO 126, expected to be published in 1981, will probably allow the option of drying at a lower temperature, such as 50°C, for those latices that are susceptible to oxidation at 70°C.

The dry rubber content of natural rubber latex is a more accurate determination than total solids content for two reasons. A much larger mass of latex is tested and the dry rubber content is less hygroscopic than the total solids content.

Dry polymer content

The dry polymer content test method is applicable to high solids 'cold' polymerised synthetic rubber latices, especially to the SBR63Y and SBR62 types. The method measures the total styrene–butadiene content, including the reinforcing polymer in the case of SBR63Y latex. It involves coagulation of approximately 6 g of latex with acetone, followed by brief refluxing and washing and drying of the coagulum to constant mass (BS 3397). Drying is carried out at 100°C. Dry polymer content is a more accurate determination than total solids content for the specified types of synthetic latex.

It should be noted that the property 'non-rubber solids' may be applied to any type of NR latex to denote the difference between total solids content and dry rubber content. On the other hand, 'non-polymer solids' is relevant only to those synthetic rubber latices to which the dry polymer content determination is applicable.

Density

Related to test methods for determination of concentration are procedures for the measurement of the density of latex. The method applicable to natural rubber latex (ISO 705) uses a 50 cm^3 density bottle and emphasises the importance of deaeration and temperature. A table of density conversion values is included for correcting the measured density to other temperatures. A second method employs a pyknometer to determine the density at 20°C of liquid plastics resins (ISO 1675). The latter procedure is, presumably, also applicable to aqueous dispersions of plastics polymers and copolymers provided they do not have a high viscosity.

ALKALINITY AND pH

In latex technology, the concept of alkalinity is particular to natural rubber latex, whereas pH is relevant to all types of latex. Alkalinity is generally preferred to pH with natural latex because it differentiates more effectively between the various types. Alkalinity, and not pH, is included in the natural latex specifications.

Alkalinity

The alkalinity of natural rubber latex is defined as the percentage by mass of ammonia (or potassium hydroxide in the case of latex preserved with that material) that it contains. It is determined by titrating latex to pH 6·0 in the presence of a stabiliser (ISO 125). Preferably the titration is carried out electrometrically but, where this is not feasible, methyl red may be used as the visual indicator. Bromothymol blue was formerly permitted as an alternative indicator until it was appreciated that it gives a lower result than methyl red. It should be noted especially that since 1977 alkalinity has been expressed on the mass of latex instead of on its water content.

pH

The pH of rubber latices is measured with a glass electrode and saturated calomel cell after standardising the pH meter with borax and potassium hydrogen phthalate solutions (ISO 976). A suitable combination electrode may, conveniently, be used in place of individual electrodes. The method gives a result that is accurate to within 0·1 pH. For aqueous dispersions of plastics polymers and copolymers, a similar method has been standardised (ISO 1148) which, however, is less rigorous in its conditions and requirements.

MECHANICAL STABILITY OF NATURAL RUBBER LATEX

Mechanical stability is an important property of natural rubber latex and a test method for its determination was first established internationally in 1957. The original procedure has been progressively modified and made more demanding, leading to the second edition of ISO 35 that is expected to be published in 1981.

The mechanical stability of natural rubber latex is defined as the time required to initiate visible flocculation by stirring the latex at high speed. The test apparatus, which is commercially available, is closely specified and the test conditions are laid down in detail. Attention is also drawn in the standard (ISO 35) to the adverse effects on mechanical stability of carbon dioxide and of lowered storage temperature of the latex sample.

It is essential that the instructions given in ISO 35 be followed in all respects in order to obtain a reliable mechanical stability result. Failure to observe any of the following conditions, in particular, will lead to an erroneous value being obtained:

(i) prevention of significant cooling of the latex between sampling and testing, especially with fresh latex,
(ii) elimination of carbon dioxide from the ammonia solution used for diluting the latex,
(iii) use of a stirrer disc of the dimensions specified (20·8 mm in diameter),
(iv) setting the clearance between the underside of the stirrer disc and the floor of the latex container at 13 mm,
(v) carrying out the determination within 24 h of first opening the sample bottle,
(vi) carrying out the determination in an atmosphere that has a normal concentration of carbon dioxide,
(vii) diluting the latex to 55·0 per cent total solids content,
(viii) filtering the diluted latex, after warming it as instructed, through the specified wire cloth (of 180 μm average aperture width) immediately before weighing out the test portion for the actual determination,
(ix) maintaining the stirring speed within 200 rpm of 14 000 rpm throughout the determination,
(x) detecting the end-point by gently spreading a sample of the latex on a suitable surface, and

(xi) confirming the end-point by over-running.

The presence of carbon dioxide in natural latex progressively and markedly lowers its mechanical stability. For example, mechanical stability is reduced by 5 per cent when the ammonia solution used for dilution contains only 0·04 per cent of carbon dioxide. A similar reduction in mechanical stability results when the carbon dioxide concentration in the atmosphere around the test apparatus is as little as 0·05 per cent. It is imperative, therefore, that the ammonia solution should be free of carbon dioxide and that the test apparatus not located in the vicinity of any equipment that generates carbon dioxide, such as certain gas or oil heaters.

The mechanical stability of natural latex is dependent on its total solids content which has been standardised at 55·0 per cent for the test. If the latex is not diluted (from 61·5 per cent total solids content), the result obtained is approximately 30 per cent lower. There is not, however, a definite and reliable relationship between the mechanical stability under standard conditions and that without dilution of the latex. If the latex is overdiluted to 50 per cent total solids content, for example, the mechanical stability is increased by approximately 50 per cent. There have been a number of attempts to alter the total solids content at which mechanical stability is measured but all have failed to gain international acceptance.

The test method insists that the latex is filtered at the latest practicable stage before it is subjected to the stirring phase. This condition is essential to ensure that there is no flocculum in the latex when stirring commences.

Maintenance of the specified stirring speed is also important. If the speed is lower than specified, an erroneously high result will be obtained, and vice-versa.

Detection of the end-point can present difficulties. The former option of taking it as the first appearance of curdiness over most of the surface of the latex is no longer acceptable since that assessment lacks reproducibility. The preferred technique is to sample the stirred latex at intervals of 15 s and to spread the sample gently on a suitable surface. The palm of the hand is particularly convenient for this purpose. The end-point is taken as the first appearance of flocculation and care needs to be taken to avoid confusion of small air bubbles with flocculum. Once flocculation has begun, it increases with continued stirring of the latex, that is with over-running, but it is the stirring time at which flocculum was first definitely detected that is the mechanical stability time of the latex. Accurate detection of the true end-point requires practice.

MECHANICAL STABILITY OF SYNTHETIC LATEX

The international test method for the determination of the mechanical stability of synthetic rubber latices is similar in concept to that for natural latex. The same stirring apparatus is used but the stirrer disc has a greater diameter (36·1 mm) and the final stage of the procedure is different. After stirring for an agreed time, the destabilised latex is filtered through 180 μm wire cloth to quantify the amount of coagulum produced.

Because stirring (at 14 000 rpm) is ineffective with more viscous latex, the test (ISO 2006) is restricted to synthetic rubber latices with a viscosity of 200 mPa.s or less. More viscous latices can be tested but only after dilution to the same viscosity range, provided that such dilution does not reduce the concentration of the latex by more than 10 per cent total solids. This limitation on dilution is set since dilution of latex decreases its stability because the balance of free and absorbed soap is changed.

The stirring time has to be agreed between 1 and 30 min to satisfy two conditions. First, the latex must not increase in temperature to more than 60°C as a result of stirring and, second, the latex must not exceed a height of 100 mm in the test container. These restrictions are designed to prevent thermal and foaming effects from dominating the test. The latex container must have a smooth inner surface to facilitate quantitative transfer of its contents after the stirring stage, and a glass container is preferred.

Some synthetic rubber latices froth excessively under the conditions allowed by ISO 2006. Certain latices with a total solids content in the vicinity of 40 per cent fall in this category. They can, however, be tested by using the smaller natural latex stirrer disc (20·8 mm diameter). In this case, it may be advantageous to limit the stirring time so that the latex does not exceed a height of 50 mm in the test container. Such conditions do not have international acceptance at the time of writing (1980) but they are, nevertheless, useful where the provisions of ISO 2006 cannot be met.

ISO 2006 specifies a high-speed mechanical stability test method. It does not necessarily indicate the stability of the synthetic latex to high shear stress, for which a rubbing test may be more appropriate. Mechanical shearing of a film of latex is the object of the Maron mechanical stability test.[1]

QUALITY OF NATURAL RUBBER LATEX

The first test method to be established as an indicator of the quality of natural rubber latex was the determination of the potassium hydroxide number (ISO 127). This test is still widely used although it is open to criticism because of the complexity of the components that it measures. More pertinent quality tests are the determinations of volatile fatty acid number (ISO 506) and carbon dioxide number.

Potassium hydroxide number
This is defined as the number of grams of potassium hydroxide that are equivalent to the acid radicals combined with ammonia in latex containing 100 g of total solids. It is determined by potentiometric titration of natural latex with potassium hydroxide solution, after adjustment of the alkalinity and dilution of the latex. The end-point of the titration is the point of inflexion of the titration curve of pH against volume of added potassium hydroxide. It occurs when the first differential reaches a maximum and the second differential changes from a positive to a negative value. Since the end-point is readily calculated, assuming linearity in its vicinity, there is no need to plot the curve. In some cases where a sharp inflexion point is not obtained, it is beneficial to repeat the titration after adjusting the alkalinity of the latex to a lower level than that specified in ISO 127.

Volatile fatty acid (VFA) number
This is a component of, and has the same units as, the potassium hydroxide number. It is determined, after coagulating the latex with ammonium sulphate and acidifying the resulting serum, by steam-distilling the serum in a Markham still and measuring the volatile acids in the distillate by titration with barium hydroxide solution. Before the titration, it is essential to free the distillate of carbon dioxide, preferably by purging with nitrogen, since carbon dioxide would also be titrated. The volatile acids in natural latex are mainly acetic acid.

Carbon dioxide number
This is also a component of, and has the same units as, the potassium hydroxide number. It is determined by either the micro-absorption method[2] or the macro-baryta technique[3] after acidifying stabilised latex. The carbon dioxide number generally has a value about 0·12 greater than the VFA number and is, therefore, an alternative indicator of the level of preservation of natural latex.

Besides measuring volatile fatty acids and carbonate and bicarbonate (collectively carbon dioxide), the potassium hydroxide number determines non-volatile acids, higher fatty acids and some other acidic materials. The fact that higher fatty acids are stabilising, whereas the other acidic components are all destabilising reduces the value of the potassium hydroxide number as a criterion of latex quality. The potassium hydroxide number adds together the destabilising and stabilising components.

Non-volatile acids (NVA)

These contribute more to the potassium hydroxide number than any of the other acidic groups. They are determined by passing latex serum, obtained by coagulation with acetic acid, through a strong cation-exchange resin, followed by evaporation and acidimetric titration.[4] The coagulation stage leaves the higher fatty acids associated with the rubber phase and converts carbonate and bicarbonate to carbon dioxide which evolves. Ion-exchange removes potassium and ammonium ions and other metal cations from the serum. Final evaporation expels added acetic acid, volatile fatty acids and any remaining carbon dioxide so that only non-volatile acids are titrated. Care needs to be taken with the evaporation, particularly in the latter stages, to ensure that all the acetic acid is expelled without thermally decomposing any of the non-volatile acids. An NVA determination occupies a total time of approximately 4 h and is a test method that requires more expertise than most. Non-volatile acids do not include higher fatty acids and are expressed quantitatively as the NVA number which has the same units as the potassium hydroxide number.

Higher fatty acids (HFA)

These are determined, after extraction of latex total solids with acetone, by acidification with sulphuric acid and working up, terminating in titration with potassium hydroxide solution.[4] Higher fatty acids are conveniently calculated as the HFA number, in the same units as the potassium hydroxide number.

Sludge content

The sludge content of natural rubber latex may also be considered as a quality criterion, although it is not widely practised. It is a property that seldom causes concern. Sludge content is determined by centrifuging the latex, repeatedly washing the resultant sludge with ammonia–alcohol solution and finally drying the sludge to constant mass (ISO 2005).

CHEMICAL STABILITY

None of the chemical stability tests proposed for latex have achieved international acceptance (in 1980), although extensive investigations have been made in this area. They have failed to meet with universal approval because of inadequate reproducibility, poor repeatability or questionable relevance to industrial latex processes.

Natural rubber latex
With natural rubber latex, stability to zinc ammine ions has been the crux of the tests that have been advocated, since an addition of zinc oxide is common to most natural latex processes. A number of tests have been suggested of which the zinc stability time (ZST) test and the zinc oxide viscosity (ZOV) test have received most favour.

In the ZST test,[5] potassium oleate solution (1·0 per cent dry on total solids) is added to the natural latex, then concentrated formaldehyde solution is carefully added to adjust the pH to 9·8 and the total solids content is reduced to 55·0 per cent with water. The temperature of the latex is adjusted to 30°C and, with mechanical stirring during a period of 10 min, 5·0 per cent of solid zinc oxide of analytical reagent grade (calculated on the total solids) is gradually added. One hour after the beginning of the zinc oxide addition, the treated latex is stirred gently, sieved and immediately a mechanical stability test is conducted in similar manner to that specified in ISO 35. The end-point is taken as the time at which flocculum is first detected in the stirred latex. A key factor in the ZST test is the use of solid zinc oxide, which was adopted to eliminate variability associated with dispersions of that chemical. It should be noted that a fatty acid soap is also added in this test and that pH is closely controlled.

In the ZOV test,[6] potassium oleate solution (1·0 per cent dry on total solids), ammonium sulphate solution (0·5 per cent dry) and zinc oxide dispersion (5·0 per cent dry) are added to the natural latex, with thorough stirring, and the total solids content is adjusted to 57·0 per cent with water. Stirring is continued for 1 min after the addition of the zinc oxide dispersion. Four minutes later the viscosity of the compounded latex is determined at 60 rpm and 25°C using the viscometer specified in ISO 1652. The ZOV test involves an ammonium salt addition to the latex, as well as fatty acid soap and zinc oxide, and it is their combined thickening effect that is measured.

Other chemical stability tests for natural latex are methods that are modifications of the ZST test or methods that use preformed zinc ammonium acetate. With the latter, one method measures mechanical stability and the other the time required for the destabilised latex to gel at an elevated temperature.

Synthetic rubber latices

Chemical stability tests proposed for synthetic rubber latices are more diverse. One method that has been considered internationally involves the addition of a surfactant and zinc oxide dispersion to the latex and measurement of the temperature and time taken to initiate coagulation. Other methods determine the proportion of coagulum formed by the addition, under specific conditions, of an electrolyte (calcium chloride) or an alcohol (methanol).

POLYMER COMPOSITION

Test methods have been standardised internationally for the determination of the bound (polymerised) styrene content and the bound acrylonitrile content of synthetic rubber latices. One bound styrene method (ISO 3136) is only applicable to SBR latices, within specified limits, and not to SBR..Y, XSBR and PSBR latices. The other bound styrene methods (ISO 4655) are more involved but they are applicable to both SBR and SBR..Y latices. The method for bound acrylonitrile content (ISO 3900) applies to (acrylo)nitrile–butadiene rubber (NBR) latices, carboxylic varieties (XNBR latices) and nitrile–isoprene rubber (NIR) latices.

Bound styrene content

ISO 3136 co-ordinates two other international standards, ISO 2028 for the preparation of dry polymer from the latex and ISO 2453 for the determination of the bound styrene content of the dry polymer. The dry polymer is prepared by coagulating the latex with sodium chloride and sulphuric acid in the presence of methanol and collecting and drying the resultant crumb. It is then extracted with ethanol–toluene azeotrope and pressed into a thin sheet. The refractive index of the sheet is measured and the bound styrene content is calculated from the result.

Total bound styrene content

ISO 4655 specifies two alternative methods for the determination of total bound styrene content and is particularly relevant to reinforced styrene–butadiene rubber latex, for which ISO 3136 is not suitable. Both methods start with coagulation of the latex with isopropanol followed by copious washing and thorough drying of the coagulum. In the carbon/hydrogen method, the coagulum is carefully combusted in a specified apparatus and the carbon dioxide and water that are produced are quantitatively absorbed. The total bound styrene content is calculated from the masses of absorbed carbon dioxide and water. This method depends on the fact that butadiene and styrene contain different proportions of carbon. In the nitration method, the coagulum is nitrated and oxidised to convert its total bound styrene content to *p*-nitrobenzoic acid, which is separated by multiple extraction and determined quantitatively by measuring its ultra-violet absorption. The carbon/hydrogen and nitration methods give comparable results for total bound styrene content when they are carried out with care.

Bound acrylonitrile content
For the determination of bound acrylonitrile content (ISO 3900), the preparative stage involves extraction of an air-dried film of the nitrile latex with water, to remove water-soluble nitrogen-containing material, and drying to constant mass. The subsequent procedure follows that for determination of the nitrogen content of natural rubber (ISO 1656). This procedure requires digestion of the prepared film with sulphuric acid, potassium sulphate and a catalyst, to convert its nitrogen content into ammonium hydrogen sulphate, from which the ammonia is distilled after making the mixture alkaline. The liberated ammonia is absorbed in boric acid solution and titrated with standard acid. The bound acrylonitrile content is calculated from the volumes of acid required in sample and blank titrations.

VOLATILE UNSATURATES

There are international test methods for the determination of volatile unsaturates in styrene–butadiene rubber latices (ISO 2008), residual acrylonitrile in nitrile latices (ISO 3899) and residual vinyl acetate in polyvinyl acetate latices (ISO 3499). Originally ISO 2008 also specified an ultra-violet spectrophotometric method for residual styrene but that method was withdrawn in 1980 since it was not sufficiently specific to styrene and was little used. The ISO 3899 method for residual acrylonitrile is sensitive only to about 0·01 per cent acrylonitrile and is now inadequate for some purposes. Consequently a much more sensitive gas chromatographic method for residual acrylonitrile is being developed, which should be standardised in the early eighties. At the same time, a gas chromatographic method for residual styrene is in the process of standardisation.

The ISO 2008 method for the determination of volatile unsaturates in styrene–butadiene latices measures other unsaturates such as butadiene dimer as well as residual styrene. The volatile unsaturates are determined by distilling the latex with methanol, adding to the distillate potassium bromate/potassium bromide solution and, after addition of potassium iodide, titrating the liberated iodine with sodium thiosulphate. The volatile unsaturates content is calculated from the volumes of thiosulphate solution required in sample and blank tests and is expressed as styrene.

Residual acrylonitrile
The ISO 3899 method for residual acrylonitrile commences with distillation of the nitrile latex and collection of the distillate in methanol. Then n-dodecyl mercaptan is added to the distillate and the excess is titrated with iodine solution. The residual acrylonitrile content is calculated from the volumes of iodine solution used in sample and blank titrations.

Residual vinyl acetate
The determination of residual vinyl acetate in polyvinyl acetate latices (ISO 3499) is similar in principle to the determination of volatile unsaturates in styrene–butadiene latices. The latex is distilled with methanol, in acid solution, potassium bromate/bromide solution is added to the distillate and the excess bromine is determined by addition of potassium iodide solution and titration of the liberated iodine with sodium thiosulphate. The result is generally expressed as the bromine number of the latex but in the case of a latex of vinyl acetate homopolymer the residual vinyl acetate content itself can be calculated.

PARTICLE PROPERTIES

Test methods for particle properties comprise determinations of viscosity, surface tension, soap content, soap deficiency and particle size. Methods for only the first two properties have been standardised internationally but various procedures are available for measurement of the others.

Viscosity
There are three international standards for the determination of viscosity, two of which use one or more models of the Brookfield viscometer. The

other standard method specifies rotational viscometers of the coaxial cylinder, cone-and-plate or double-cone types. One method applies to rubber latices and the other two to dispersions of resins or polymers.

ISO 1652 describes the determination of the viscosity of rubber latex. It specifies the Brookfield L instrument for low viscosities (up to 200 mPa.s), the R instrument for high viscosities (above 2000 mPa.s) and permits either instrument to be used for intermediate viscosities. The speed of rotation of the viscometer is restricted to 60 rpm for the L instrument and 20 rpm for the R instrument. In principle, this viscometer measures the torque produced on a specified spindle rotating at constant speed and a low rate of shear while immersed to a known depth in the latex. The scale reading that is obtained is easily translated into the viscosity in millipascal seconds. In the viscosity range 50 to 200 mPa.s, the R viscometer operated at 50 rpm gives similar results to the L instrument (at 60 rpm), but it is only an approximate equivalence that is not recognised by the international standard.

ISO 2555, for the determination of the viscosity of dispersions of resins, differs from ISO 1652 in two respects. Only the R model of the Brookfield viscometer is specified and there is no restriction on the speed of rotation, from 0.5 to 100 rpm, although speeds of 10 and 20 rpm are recommended.

ISO 3219 is a general standard, for the determination of the viscosity of dispersions of polymers, that uses a rotational viscometer working at a defined shear rate. Seven shear rates from 1 to 250 reciprocal seconds (or these values multiplied or divided by 100) are recommended. The viscosity results are expressed in pascal seconds, qualified by the shear rate employed.

Since the viscosity of latex increases with its total solids content, as well as depending on the rate of shear used, both total solids content and shear rate should be appended to all viscosity results.

Surface tension
Determination of the surface tension of rubber latex is the subject of ISO 1409. This standard specifies the du Nouy tensiometer with a platinum ring of either 60 mm or 40 mm nominal circumference. The latex is tested at a total solids content of 40 per cent or less, since at greater concentrations the higher viscosity of the latex may affect the accuracy of the determination. In France, a thin rectangular platinum plate is specified as an alternative to a ring for the measurement of surface tension. The units of surface tension are millinewtons per metre (mN/m). Surface free energy is synonymous and has

the same numerical value as surface tension when it is expressed in millijoules per metre squared (mJ/m^2).

Soap content

The soap content of synthetic latices that have been produced with fatty acid soap or rosin acid soap can be determined by potentiometric titration after stabilisation of the latex. One method that is applicable where the soap in the latex is potassium oleate begins by diluting latex with water and adding equal volumes of a non-ionic stabiliser and neutralised isopropanol followed by potassium hydroxide to increase the pH to at least 11·0. The treated latex is titrated potentiometrically with sulphuric acid, through two inflexion points that occur at pH values of approximately 9·4 and 5·0. The inflexion points can be determined in the manner of the potassium hydroxide number method (ISO 127). The volume of acid corresponding to the pH interval between the two inflexion points is calculated for both a sample titration and a blank titration, and the potassium oleate content of the latex is calculated from the difference between these two volumes.

Soap deficiency

The soap deficiency of some latices can be determined by surface tension titration or conductimetric titration with the soap that is already present in the latex. These methods are usually appropriate where the soap is potassium oleate. In the case of the surface tension approach, incremental addition of soap to a latex deficient in soap lowers the surface tension more or less linearly until a point is reached after which further addition of soap reduces the surface tension only slightly. The break in the graph of surface tension against soap addition represents the amount of added soap at which the polymer surface is saturated with adsorbed soap. This amount of added soap is the soap deficiency of the latex. With conductimetric titration, electrical conductance of the latex increases linearly with the soap addition, rapidly at first and then more gradually. The break in the graph indicates the soap deficiency.

Under favourable circumstances, a soap surplus in a latex can be determined by blending it with an excess of a compatible latex of known soap deficiency and subjecting the blend to surface tension or conductimetric titration. The soap must be the same in the two latices and their proportions must be such as to produce an overall soap deficiency. Blending has to be carried out carefully and the blend allowed to come to equilibrium before it is tested.

Particle size

There are several established methods for investigating the particle size of latex, although none have become international standards. Techniques that are used include electron-microscopy, soap adsorption, light-scattering, centrifugation, fractional creaming and counting, methods. Electron-microscopy is most readily applicable where the distribution of particle size is relatively narrow and the particles do not flatten appreciably during drying of the latex sample on the support film. Rubbery polymers need to be hardened by bromination or some other means to prevent coalescence or distortion of the particles. Latices with a wide particle size distribution, such as high-solids rubber latices, can be analysed by electron-microscopy but a large number of particles have to be measured in order to obtain useful results.

Soap adsorption methods, developed by Maron et al.,[7] require the latex to have a soap deficiency, which is determined as described above. The soap content and the molecular adsorption area of the soap need to be known. Assuming that the latex particles are geometrically simple, the specific surface area of the latex and hence an average particle size can be calculated. Light-scattering methods for particle size determination measure the intensity of radiation that is scattered by the latex at various angles to the direction of incidence. Centrifuge methods depend on the difference in density between the particles and the serum in which they are dispersed. The Coulter Counter is a useful instrument for measuring the size distribution of larger latex particles and a more recent model (in 1980) rapidly and automatically measures average particle size in the range 40 to 3000 nm.

The fractional creaming technique for the determination of latex particle size distributions was developed by Schmidt and Biddison.[8] This method exploits the quantitative inverse relationship between concentration of a creaming agent (sodium alginate) and the size of creamed particles. By preparing a calibration curve with the aid of an electron-microscope, particle size distributions of latices over a range from 50 to 1000 nm diameter can be determined with only an analytical balance and a number of separating funnels.

A useful comparative indicator of the particle size of latex, that can be obtained quickly, is the limiting solids concept. This is the total solids content at which the latex has a specified viscosity. A higher total solids content indicates a larger particle size, and a lower solids content points to a smaller particle size. Alternatively, the viscosity at which the latex has a specified total solids content can be determined.

CHEMICAL TESTS

There are international standards for the determination of the copper content (ISO 1654), manganese content (ISO 1655), iron content (ISO 1657) and nitrogen content (ISO 1656) of rubber latex. The first three test methods apply to both natural and synthetic rubber latex and the method for nitrogen is designed for natural latex. All four standards require the latex to be converted to its total solids before analysis, and those for copper, manganese and iron involve an ashing stage. Copper is determined by reaction with zinc diethyl dithiocarbamate and the measurement of optical density. Manganese is determined by the potassium periodate photometric method and iron by the 1,10-phenanthroline photometric method. An outline of the nitrogen determination is given in the last paragraph of the section devoted to polymer composition. The nitrogen content of natural rubber latex may be used to obtain an estimate of its protein content, by applying a factor of 6·25.

Recent developments in flameless atomic absorption spectroscopy make possible the direct determination of copper, manganese and iron in diluted latex, without the lengthy, conventional ashing stage. This technique can also be applied to the determination of a number of other elements in latex.

OTHER TEST METHODS

There are three other international standards for latex, to which attention should be drawn. One relates to natural rubber latex and the other two to plastics latices.

ISO 1147 specifies a freeze–thaw cycle stability test for plastics latices, in which the latex is subjected to five cycles of 16 h at $-10°C$ followed by 8 h at room temperature. The method may also be applied to rubber latex, although it has been criticised on the grounds that the scale of the test is too small for its results to be translatable to commercial situations.

ISO 1802 describes the determination of the boric acid content of the LABA type of natural rubber latex. The test method involves complexing the boric acid with mannitol and titrating the liberated hydrogen ions with alkali.

ISO 2115 concerns the determination of the white point temperature and minimum film-forming temperature of plastics latices. The method entails application of a temperature gradient to a suitable metal plate, spreading a thin film of the latex on the plate, drying it and determining the temperature at which the coalesced (transparent) section of the latex meets the uncoalesced (white) section. The minimum film-forming temperature is the lowest temperature at which a homogeneous film is formed.

SUMMARY

This chapter shows that the range of international standards for latex is comprehensive and that several other test methods have been developed. The published standards are subject to periodic review and some of them may be further refined in the future. It is unlikely that there will be many new developments in the field of latex testing, although commercial availability of new types of latex may necessitate the standardisation of specific test methods.

ACKNOWLEDGEMENT

Quotations in this chapter from British and International Standards are reproduced by permission of the British Standards Institution, 2 Park Street, London, WIA 2BS, England, from whom the complete Standards can be obtained.

REFERENCES

1. MARON, S. H. and ULEVITCH, I. N., *Anal. Chem.*, 1953, **25**, 1087.
2. CALVERT, K. O. and SMITH, R. K., *Rubb. Ind.*, 1974, **8**(1), 31.
3. SUNDARAM, P. and CALVERT, K. O., *Proc. Int. Rubb. Conf.*, Kuala Lumpur, 1975, 314.
4. CALVERT, K. O., *Plastics and Rubber: Materials and Applications*, May 1977, 59.
5. NEWNHAM, J. L. M., CALVERT, K. O. and SIMCOX, D. J., *Proc. Nat. Rubb. Res. Conf.*, Kuala Lumpur, 1961, 668.
6. DAWSON, H. G., *Rubb. World*, 1956, **135**, 239.
7. MARON, S. H., ELDER, M. E., ULEVITCH, I. N. and MOORE, C., *J. Coll. Sci.*, 1954, **9**, 89, 104, 263, 347, 353 and 382.
8. SCHMIDT, E. and BIDDISON, P. H., *Rubb. Age, NY*, 1960, **88**, 484.

APPENDIX: INTERNATIONAL AND NATIONAL STANDARDS CITED

(Note: The wording of the titles of the international standards listed below may be modified but their ISO numbers will not be altered.)

ISO 35: Rubber latex, natural — Determination of mechanical stability.
ISO 123: Rubber latex — Sampling.
ISO 124: Rubber latices — Determination of total solids content.
ISO 125: Rubber — Natural latex — Determination of alkalinity.
ISO 126: Rubber latex, natural — Determination of dry rubber content.
ISO 127: Natural rubber latex — Determination of the potassium hydroxide number.
ISO 506: Natural rubber latex — Determination of volatile fatty acid number.
ISO 705: Natural rubber latex — Determination of density.
ISO 706: Rubber latices — Determination of coagulum content.
ISO 976: Rubber latices — Determination of pH.
ISO 1147: Plastics — Aqueous dispersions of polymers and copolymers — Freeze-thaw cycle stability test.
ISO 1148: Plastics — Aqueous dispersions of polymers and copolymers — Determination of pH.
ISO 1409: Rubber latex — Determination of surface tension.
ISO 1625: Plastics — Aqueous dispersions of polymers and copolymers — Determination of residue at 105°C.
ISO 1629: Rubbers and latices — Nomenclature.
ISO 1652: Rubber latex — Determination of viscosity.
ISO 1654: Raw rubber and rubber latex — Determination of copper.
ISO 1655: Raw rubber and rubber latex — Determination of manganese content — Potassium periodate photometric method.
ISO 1656: Raw natural rubber and natural rubber latex — Determination of nitrogen.
ISO 1657: Raw rubber and rubber latex — Determination of iron content — 1,10-Phenanthroline photometric method.
ISO 1675: Plastics — Liquid resins — Determination of density by the pyknometer method.
ISO 1802: Natural rubber latex — Determination of boric acid.

ISO 2004: Rubber latex, natural — Centrifuged or creamed, ammonia-preserved types — Specification.
ISO 2005: Natural rubber latex — Determination of sludge content.
ISO 2006: Synthetic rubber latex — Determination of high-speed mechanical stability.
ISO 2008: Rubber latex, styrene–butadiene — Determination of volatile unsaturates.
ISO 2027: Rubber, natural latex, evaporated, preserved — Specification.
ISO 2028: Butadiene homopolymer and copolymer latices — Preparation of dry polymer.
ISO 2115: Plastics — Aqueous dispersions of polymers and copolymers — Determination of white point temperature and minimum film-forming temperature.
ISO 2438: Rubber latices, synthetic — Codification.
ISO 2453: Styrene–butadiene copolymers — Determination of bound styrene content.
ISO 2555: Resins in the liquid state or as emulsions or dispersions — Determination of Brookfield RV viscosity.
ISO 3136: Rubber latex, styrene–butadiene — Determination of bound styrene content.
ISO 3219: Plastics — Polymers in the liquid, emulsified or dispersed state — Determination of viscosity with a rotational viscometer working at defined shear rate.
ISO 3499: Plastics — Aqueous dispersions of homopolymers and copolymers of vinyl acetate — Determination of bromine number.
ISO 3899: Rubber — Nitrile latex — Determination of residual acrylonitrile content.
ISO 3900: Rubber — Nitrile latex — Determination of bound acrylonitrile content.
ISO 4576: Plastics — Aqueous dispersions of polymers and copolymers — Determination of gross particle content by sieve analysis.
ISO 4655: Rubber latex, reinforced styrene–butadiene — Determination of total bound styrene content — Carbon/hydrogen method and nitration method.
BS 3397: Methods of test for synthetic rubber latices.
BS 4661: Emulsion polymerised anionic styrene–butadiene rubber latices.

Chapter 5

LATEX APPLICATIONS IN CARPETS

D. PORTER

Lintafoam (Manchester) Ltd, Rossendale, UK

One of the largest users of latex is unquestionably the carpet industry, in particular the tufted carpet sector which has shown rapid growth since its conception in the early 1950s. The 1978 market statistics[1] depicted in Table 5.1 for the three carpet manufacturing techniques (tufted, woven and needlefelt) in the major carpet producing countries clearly demonstrates the dominant role occupied by the tufting machine in the production of broadloom carpet. Equally dominant is the US tufting industry, which produces more than all other countries added together, at an annual production figure of 866 million square metres.

Latex, in its many forms, is used for a variety of important application processes within the tufted carpet industry. The three main techniques are precoating, secondary backing and foam backing. In briefly describing these three techniques, precoating is the application of a carboxylic-styrene–butadiene (XSBR) latex compound to lock in the tufts, secon-

TABLE 5.1
Carpet volumes of the major manufacturing countries

		Million m^2	
	Tufted	Non-woven	Woven
United States	866·0	17·0	12·5
Germany	125·0	32·0	9·0
United Kingdom	132·0	6·0	39·0
Belgium	110·0	24·0	26·0
Canada	72·0	6·0	4·0
Netherlands	35·0	8·5	2·5
France	26·5	23·0	1·7
Denmark/Sweden	25·5	1·4	1·7

Source: Intercontuft, 1978.[1]

TABLE 5.2
Latex volumes used in the various backing systems

	Foam	Precoat	Secondary backing
	\multicolumn{3}{c}{1 000 tons}	
United States	26·0	9·0	120·0
Germany	32·0	9·0	2·4
United Kingdom	26·0	8·0	8·0
Belgium	21·0	7·0	3·3
Canada	10·0	3·5	8·5
Netherlands	8·0	3·0	Small
France	7·0	2·4	Small
Denmark/Sweden	6·3	2·0	Small
Total	136·3	43·9	142·2

Source: Authors calculations based on carpet volumes detailed in Table 5.1; combined with estimates of filler levels and applied weights for each system.

dary backing also involves an XSBR latex compound but this time as an adhesive for lamination of a secondary fabric and foam backing is the application of an SBR latex, natural latex (or a blend of the two), to form a thin cushion which acts as an integral foam underlay. The volume of latex utilised in conjunction with tufted carpet is not easily derived from statistical data but an approximation can be made given the carpet volumes detailed in Table 5.1 and by estimating the rough averages for applied weight and filler levels applicable for each listed country's carpet statistics. These calculated figures are given in Table 5.2. Probably the most interesting statistic is the large volume of XSBR which, if precoat and secondary backing volumes are combined, is considerably greater than that used for foam backing. The reason is easily explained and is due to the influence of the US markets preference for secondary backing in contrast to the European industry where each individual national market enjoys anywhere from 70 to 90 per cent foam backing.

PRECOATING

Carpet direct from a tufting machine is limp and virtually unusable as a floor covering without further processing. It is essential to apply some form of coating for stabilisation and to anchor the individual tufts firmly to the primary tufting cloth. This process was originally called anchorcoating but it is now given the more universal term of precoating.

TABLE 5.3
Precoat formulations for tufted carpets

	Per cent activity	Parts by mass (dry) (a) Lick coating	Parts by mass (dry) (b) Froth coating
Carboxylic SBR latex (50–60 per cent styrene)	52	100	—
Carboxylic SBR latex (60–70 per cent styrene)	52	—	100
Sodium lauryl sulphate	28	—	0·75
Sodium hexametaphosphate	25	0·50	0·50
Defoamer	100	0·50	—
Calcium carbonate filler	100	600	500
Polyacrylate thickener	16	to viscosity 5–10 Pa.s	—
Water	—	to 78 per cent TSC	—

Precoat compounds are simple mixtures of XSBR latex and an extender, which is almost always a ground limestone or whiting filler. Table 5.3 illustrates the broad spectrum of formulations used. Formulation (a) is probably the most typical and is applied by the conventional 'lick roll' technique. The styrene content of the latex polymer ranges from 50 to 65 per cent depending on whether a soft or stiff 'hand' is required. For high tuft lock the higher styrene level is needed and the accompanying rigidity must, therefore, be accepted. A small amount of polyphosphate is advisable with most latices in order to sequester free calcium ions and maintain long-term viscosity stability. Polyphosphates also aid filler dispersion, particularly when large quantities are involved. Defoamers are sometimes used to reduce the tendency for frothing in the lick roll bath. Frothing occurs to different extents across the width of the bath, causing areas of low specific gravity which when eventually applied to the carpet gives areas of underweight application perhaps sufficient to result in a soft and unacceptable 'hand'. Filler is added in substantial quantities and is added dry; it is not predispersed. Its inclusion is not just for cost reasons because it does have some technical benefits. For example, it allows the application of a high weight of what has now become a cheap compound to provide the stiffness and body required of the tufted carpet. To control penetration into the tufts and to minimise filler settling out of suspension, the compound viscosity is increased by the use of thickeners such as polyacrylates.

Instead of applying the precoat in the form of a high specific gravity liquid, it has become increasingly popular to 'froth' the compound before it enters the lick bath, principally to increase the volume by reducing specific gravity. In so doing, it permits a reduction in application weight combined with easier control.

Formulating is only slightly modified (see Table 5.3 formulation (b)) and involves the stiffer polymer to compensate for the reduced weight and inferior rigidity of a reticulated film. Frothing is aided by the addition of a small amount of sodium lauryl sulphate surfactant and obviously the need for a defoamer is no longer necessary. Filler levels are slightly lower than for straight 'lick' coats at 400–500 phr to ensure satisfactory film strengths are maintained.

Applying the compound to the carpet is one of the most simple operations involved in tufted carpet manufacture and utilises the lick roll and scraper blade principle (Fig. 5.1). As the lick roller revolves, either with or against the carpet direction, it transfers the compound onto the back of the carpet. The excess is removed by the scraper blade which also assists in forcing and controlling penetration into and around the individual tufts. It should also be said that there are other factors which also control the degree of penetration such as compound viscosity, direction and speed of the lick roller, carpet contact angle across the apex of the roller and carpet tension. When first setting up a new machine, all these factors need to be considered but, once established, further day-to-day control tends to concentrate on knife angle and lick roll speed. Finally, the carpet is dried either by infra-red heaters directed at the precoat side only or in a short oven where drying takes place from

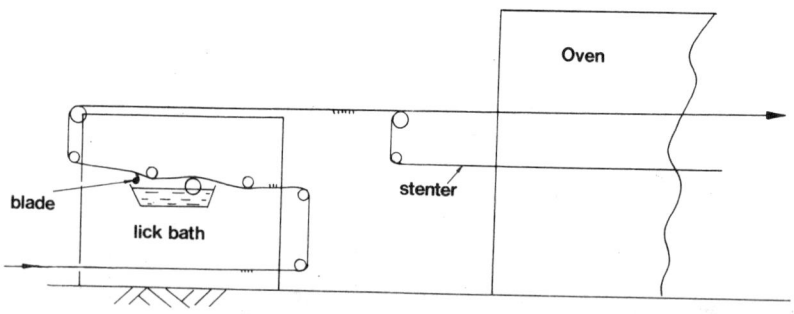

Fig. 5.1. Precoating of tufted carpets.

both sides of the carpet. The latter is the most efficient and controllable drying process.

Already a brief mention of the precoat's function has been made. However, it is worth further elucidation because for such a crude latex compound it performs one of the most vital functions of any latex compound applied to tufted carpets. In the first place its primary task is to lock the tufts into the primary scrim so that to remove them requires a significant force, i.e. approximately 1·5 to 2·0 kg/tuft for cut pile and 3 to 5 kg/tuft for loop pile. Extra weight is another factor. The average pile weight of European tufted carpet is as little as 440 g/m^2 and with applied weights usually between 550 and 650 g/m^2 precoats often double the weight of most carpets. Dimensional stability is the last of the most important properties conferred by the precoat and it achieves this by preventing distortion during processing and subsequent service.

Precoat formulations are easily modified to impart special features required for some critical installations. Three typical examples are given below.

(a) Carpets for hospital use require spillages (particularly in geriatric wards) to remain within the pile surface and not to penetrate through the precoat barrier. This requires a precoat which has been made water repellent by using either a purpose designed XSBR latex or a hydrophobic polymeric additive.

(b) Often a carpet may have to conform to certain flammability specifications. In fact, passing the test[2] is mandatory for all carpets sold in North America, whether domestic or contract, whilst in Europe flammability requirements are only specified in special circumstances. However, regardless of which flame specification is quoted, or its degree of severity, flame retardancy is imparted by replacing some or all of the calcium carbonate filler with alumina trihydrate (ATH). Typically 200 phr ATH is normally sufficient to pass the pill and hot nut test,[4] whereas 400–500 phr may be required to pass a radiant panel test.[3]

(c) Loop pile tufted carpets can easily be inadvertently or deliberately snagged and because of their construction a whole row of tufts will be removed. Defects of this nature are usually referred to as 'ladders' and are a problem common in schools where pencils are used to lever out the first few tufts and thereafter the whole row. To prevent such 'accidents' precoats capable of producing a 10 kg tuft lock are being specified. Compounds to meet this high standard are formulated using an XSBR polymer (or blend) having a styrene content of 65–70 per cent and are compounded with only 200–300 phr filler. Whilst this will provide good tuft lock, an applied weight of 900–1000 g/m² will be required to achieve the 10 kg.

Primary backings
Primary backings, now more commonly referred to as unitary backings, are very similar in concept to precoats. They involve the same latices and chemicals and are applied to carpets in exactly the same way. The only difference is in the filler level. Unitary backings usually only contain 200–300 phr and as a consequence the coating is more flexible and, therefore, resists cracking on flexing. They are used, as their name implies, as a single latex treatment and remain visible to the end-user, unlike precoats which are subsequently covered by a foam. This explains the need for a strong and flexible film but it is also because unitary backings are often used for direct gluedown contract installations where the film strength must be sufficient to allow the worn carpet to be removed from the floor without leaving behind a time-consuming and expensive clean-up operation.

SECONDARY BACKINGS

Most people associated with the carpet industry agree that secondary backing, also termed jute backing, was not invented for any specific technical reason. It came about because the consumer identification between tufted and traditional woven carpets was readily visible. To mask the difference a secondary jute fabric was laminated to the back of the tufted carpet in an attempt to imitate woven carpets such as Wiltons and Axminsters.

The original techniques were based on natural latex, using particularly its auto-adhesive qualities, but over the years it has developed towards specially designed XSBRs. Today the process involves latex coating the back of the carpet, in exactly the same manner as for precoats, and laminating to it the secondary jute before drying in an air circulating oven (Fig. 5.2). To produce a satisfactory end product the adhesive qualities of the compound must be good, but that alone is insufficient to guarantee success. It is just as important, perhaps even more so, to ensure that the two substrates are in contact throughout the drying operation. There must be no point in the process, particularly during the early stages, where unequal tensions create stress, otherwise the bond will

LATEX APPLICATIONS IN CARPETS

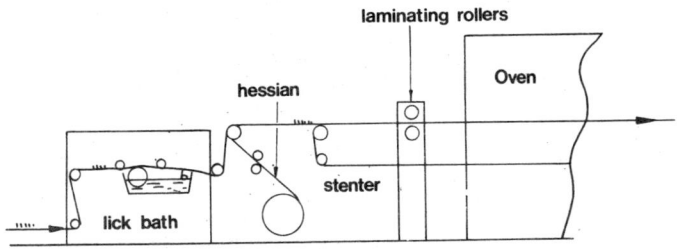

FIG. 5.2. Secondary backing of tufted carpets.

be weakened or destroyed altogether regardless of the adhesive qualities of the compound. To minimise the risk of this occurring, the carpet is processed 'pile up' so that the carpet weight acts downwards towards the adhesive and hessian.

Table 5.4 outlines a typical formulation recipe. The XSBR latex is designed specifically to exhibit a high degree of wet tack by polymerising with low surfactant levels. However, it exhibits little or no tack in the dried film state which emphasises how critical the early stages of the process are. Styrene levels are normally in the 55–60 per cent region to ensure a strong ultimate bond and total product.

TABLE 5.4
Secondary backing formulations for tufted carpets

	Per cent activity	Parts by mass (dry)
Conventional Lick Coating Technique		
Carboxylic SBR latex (55–60 per cent styrene)	52	100
Water	—	to 78 per cent TSC
Sodium hexametaphosphate	25	0·25
Calcium carbonate filler	100	300
Polyacrylate thickener	16	to 10 Pa.s
High Solids Froth Technique		
Carboxylic SBR latex (specially designed)	54	100
Water	—	to 82–84 per cent TSC
Sodium hexametaphosphate	25	0·5
Sodium lauryl sulphate	28	0·5
Calcium carbonate filler	100	400
Thickener (polyacrylate or cellulosic)	16	to 20 Pa.s

The essential feature of compounding is not to destroy the polymers inherent wet tack properties by, for example, adding unnecessary surfactants or dispersing agents, excessive quantities of filler or by reducing the compound solids with excessive water. This last point is particularly important since a low solids compound requires excessive thickener levels to obtain a workable viscosity and then the combination of the two creates a slow drying compound which delays the onset of the critical wet tack stage. Taking all these considerations into account, the compounds listed in Table 5.4 contain no added surfactant; only a small quantity of polyphosphate is added; filler is kept to a moderate level of 300 phr and is often of fine particle size to promote a high viscosity compound which, therefore, requires only minimal thickener addition. A compound made to this recipe should display a high degree of wet tack which will rapidly develop and dry.

Other important considerations related to successful secondary backing are the fabric type and the applied compound weight. In the case of the former, hessian is the most universally used secondary fabric at a weight of 200–240 g/m². Other synthetic fabrics made from polyester and polypropylene are gaining ground, particularly the latter and especially in North America. Compound application weights vary according to the carpet type, the determining factor being total coverage of the backstitch. Generally though the applied weight lies between 800 and 1100 g/m².

High solids froth secondary backing

A modified version of secondary backing was developed in 1976,[5] and described as a high solids froth secondary backing process. It is based on the principle that if the total solids can be raised then the development of the critical wet tack stage is achieved earlier thus giving a faster, safer system. However, it is not quite that simple because there are several inter-related factors which contribute to the ultimate success of the system.

To begin with the total solids aspect; there are only two ways to obtain a high total solids, either by having a latex of high solids or by increasing the filler level. For practical purposes only a slight increase from 50 to 54 per cent in latex solids is possible but this alone is insufficient. Raising the filler level from the recognised standard of 300 phr to 400 phr is the easiest and cheapest solution. The next important factor is the polymers ability to readily accept filler at this high level without added water and without producing a viscosity problem. To do so usually means a special secondary backing XSBR latex which can readily accept and disperse the filler to give a compound displaying good viscosity stability on storage. The third factor is that a compound containing more filler will theoretically exhibit less wet tack, but this is compensated for in two ways. First, the higher solids allows faster development of tack even though it is not as great and, second, by building in improved compound rheology through the dual mechanism of a higher viscosity compound and then further increasing it by frothing to a specific gravity of 1·0 (originally 1·6). This ensures that the compound is relatively thixotropic and hence remains on the peaks of the tufts. Systems displaying this characteristic are referred to as 'high riding' compounds. The last important factor is connected with aeration, or frothing, which, because it provides more volume for a given weight, helps to fill in the valleys between the rows of tufts and, therefore, provides more compound in contact with the hessian. In some instances it is even possible to reduce the applied weight.

A typical formulation for the 'high solids froth' is given in Table 5.4 (formulation 2). The XSBR latex is as described in the previous paragraph, to which is added 400 phr of calcium carbonate or ATH if a flammability specification is necessary. A polyphosphate is included as is the standard practice, along with a lauryl sulphate surfactant to produce adequate foamability and foam stability. Finally, the compound is brought to the desired viscosity using a polyacrylate or cellulose thickener.

Processing is the same as for conventional secondary backing with just one exception — *en route* to the lick bath the compound is aerated in an automatic foamer. Good mixing of the air is important if a high foam stability is to be achieved. Coarse foam will rapidly break down in the bath resulting in uneven weight application. Cellulose thickeners, which aid foam stability, are an added advantage from this point of view.

Advantages claimed for the 'high solids froth' technique are extremely attractive and include:

(i) Reduced costs — by increasing the filler level there is automatically a reduction in compound raw material cost. Additional raw material savings can be made if the improved covering power of the system also permits a weight reduction.

(ii) Improved safety margins — due to the combination of high solids and hence early achievement of bond, plus its 'high riding' characteristics, there is less risk of bond disturbance during the early critical stages of the bonding process.

(iii) Faster running speeds — a factor which is entirely due to the fact that there is considerably less water to evaporate. Speed increases of 20–40 per cent are typical but depend entirely on the solids content of the original conventional system.
(iv) Labour and energy savings — obviously if the process is capable of extra speed then the carpet volume manufactured per unit time is greater. Consequently, labour and energy costs are less per unit area of finished product.

Physical properties derived from secondary backing
Carpet properties such as tuft lock, pilling and fuzzing resistance, added weight and body are automatically served by the latex compound. The benefit from the secondary cloth is two-fold; first, it satisfies the aesthetic aspect associated with promoting a quality image which accounts for why secondary backed carpets are directed at the middle to top end of the market and, second, it confers enhanced dimensional stability and 'hand'.

FOAM BACKING

Foam backing of tufted carpets started in the USA and was first adopted in Europe in 1959[6] using a narrow width fabric coating machine. From that day up until the present time, equipment and techniques have been dramatically improved to such an extent that it is now common place to foam back carpet (5 m wide) at speeds of 10 m/min. Initially foam application involved a gelation step using either the zinc ammine or the sodium silicoflouride processes, but by about 1969 a new non-gel process was becoming increasingly popular and by the middle of the 1970s it had become the most used technique for foam backing tufted carpets.

Foam application technique
In order to understand better the salient points of succeeding sections associated with foam backing, the technique of how a foam is applied to carpet is best explained first.
 Regardless of the type of foam system being applied, foaming the latex compound, spreading onto carpet and the subsequent vulcanisation and drying is the same virtually the world over. To begin with, it is essential

wherever a foam is used to apply a precoat first (see Fig. 5.3). Once the precoat is substantially dry the foam can then be applied. It is done so from an automatic foaming machine which serves to pump compound from the holding tank through the mixing head, where air is homogeneously dispersed throughout, before being fed to the foaming station by means of a long reinforced plastic pipe. It is deposited evenly across the carpet width (which can be as wide as 5 m) using a traversing trolley situated above the doctor roller. A foam reservoir is maintained between the doctor roller and the deckles, the latter also controlling the coating width. Foam gauge is controlled by the doctor roller (some plants are equipped with blades or a knife) which has the facility to be adjusted independently at both ends.

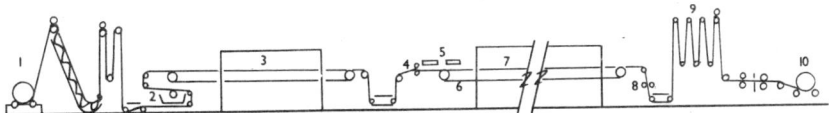

FIG. 5.3. In-line precoat and foaming plant. 1, let off; 2, precoat application; 3, precoat drying oven/pre-stenter; 4, foam application; 5, infra-red zone; 6, main stenter; 7, foam drying oven; 8, trimming; 9, festoon; 10, batch up.

After foam application the carpet is automatically carried onto a pin stenter which, as well as carrying the carpet through the oven, also prevents carpet shrinkage during drying and vulcanising. The first 2–3 m of the stenter (termed the 'wings') is narrower than the rest which permits easy pinning on and then a steadily increasing lateral tension is created as the wing section gradually becomes wider to meet the full width of the stenter.

Situated directly after the 'wings' is usually the infra-red zone which provides the necessary heat for gelation or, in the case of non-gels, to skin the foam surface. From there the carpet passes into the drying and vulcanising oven which today is usually a direct gas fired modular oven of 3-m sections with the facility to operate at lower temperatures on the pile side, an innovation termed 'split zoning', and designed to allow the processing of temperature sensitive fibres like polypropylene and acrylics.

Finally, the product leaves the oven to enter a cooling zone of high velocity air before being released from the stenter. It is then trimmed to size, batched in desirable roll lengths (typically 25 m) and wrapped ready for despatch.

The zinc ammine gelation system

The zinc ammine heat sensitive gelation system or, as it is known in the carpet sector, the ammonium acetate system, was used long before foam carpet backing became popular but was ideally suited for that purpose. It is a complex chemical reaction involving zinc ions (derived from zinc oxide), an ammonium salt (usually the acetate) and ammonia. Its basic chemistry has been well documented[7] and can be represented by the following mechanism:

$$Zn^{++} + 4NH_3 + NH_4^+ \rightarrow Zn(NH_3)_4^{++}$$

$$RCOO-Zn-OOCR \xleftarrow[\text{(fatty acid soap)}]{2RCOO^-} Zn^{++} \updownarrow$$

Gelation of latex involves a phase inversion from rubber in water to water in rubber and for it to be achieved the protective colloid must be destroyed. For most latices the colloid stability is attributable to a fatty acid soap which, during the gelling step, is converted from a soluble and ionised soap to an insoluble and non-ionised zinc soap no longer capable of conferring colloidal stability to the latex particles. The latex particles can then coalesce to form a continuous gelled mass of polymer. It is postulated that the first stage of the reaction is the formation of the zinc tetrammine complex which has little effect on latex stability. However, this complex is unstable and, by heating, the ammonia volatilises off forming the lower order complexes and ultimately the free zinc ion. Gelation is ultimately achieved through the free zinc ion reacting with the soap on the latex particles and forming the insoluble zinc soap.

Putting this gelation technique into practice on a carpet plant is not difficult but unfortunately it does not enjoy a strong position today following the advent of non-gel technology. This prejudice is connected with the simplicity of the non-gel system whereas producing a good product from the zinc ammine system is much more dependent on factors such as correct gel addition levels, quality of chemicals or intensity of heat applied, any of which may be the cause of substandard product if incorrectly set. For these reasons the major carpet markets only use the zinc ammine process when the physical property requirements demand a gel foam.

TABLE 5.5
Zinc ammine gelled foam recipes

	Per cent activity	Parts by mass (dry) 1	2
Styrene–butadiene latex	66	50	100
Potassium oleate	20	3·0	4·0
Water	—	to 70 per cent TSC	to 72 per cent TSC
Natural rubber latex	61	50	—
Sulphur	50	2·25	2·0
ZDEC	50	1·25	1·0
ZMBT	50	1·00	1·0
Antioxidant[a]	40	1·00	1·0
Zinc oxide	50	3·0	3·0
Calcium carbonate filler	100	100	120
Polysiloxane emulsion	50	0·15	0·2
Polyacrylate thickener	16	0·10	0·10
Ammonium acetate	20	2·0	3·0

[a] In practice the antioxidant type is usually one (or a synergistic pair) designed for heat protection and not light. Therefore, the tendency is to use quinoline-based or hindered phenolic antioxidants.

Two typical formulation recipes are given in Table 5.5. The first is designed for high quality flat and waffle foams. The combination of high natural latex and low filler level contribute to a wide gel range, allowing easier processing by virtue of its freedom from gel cracks and foam collapse. For stabilisation and foamability a soap is required which must also be capable of destabilisation through the zinc ammine, gelation mechanism. Normally a fatty acid soap like potassium oleate serves this purpose well. The cure system of sulphur, ZDEC, ZMBT, and zinc oxide is typical of flat gelled foam and is designed to give a cure time of approximately 10 min at 150°C for an application weight of about 800 g/m^2 at 3 mm gauge. A filler level of 100 phr is modest and, like almost all latex carpet applications, is a calcium carbonate derived from whiting or limestone deposits. There are no hard and fast rules related to filler particle size, principally because the type used is cheap which means that it is obtained from local deposits and, of course, the quality will vary from area to area. However, fillers containing high levels of trace metals, such as copper and manganese, must be avoided if the foam is to have good ageing resistance. Equally the extremes of coarse and ultra fine particles are not conducive to either aesthetic appearance or easy processing. Nonetheless, the system tolerates a wide spectrum of differing particle size filler. An annoying feature of the zinc ammine process is the flattening of the cellular structure and the layering of the cells. As a consequence of this cellular stratification, optimum physical properties (particularly tensile and elongation) are difficult to achieve. Furthermore, it causes a weak interface between foam and precoat resulting in a peel bond (often termed delamination strength) which is inadequate for the function it has to perform. A small quantity of a polysiloxane emulsion[8] overcomes this defect by refining the foam structure to such an extent that all physical properties are radically improved. Nowadays a silicone is a prerequisite for any zinc ammine gelled foam applied to carpet. Without it the system would find only limited use in the carpet area.

As the industry has gained knowledge and skills, the trend has been towards cheaper formulations (Table 5.5, formulation 2) based on 100 per cent SBR latex. For vertical carpet mills this has the advantage of the need to stock only one foam latex. Compounding ingredients are exactly the same as in the expensive formulation with the curatives and antioxidants being supplied from a compounder in the form of a cure paste which further simplifies matters for the carpet mill. Provided the carpet quality permits a low gauge application which is flat and free from projections through the foam surface, as much as 120 phr of filler can be used. An example would be spreading at 2·5 to 3·0 mm at a density of 350 g/litre on a low loop pile carpet for contract use which is typical of many continental European products. Lower filler levels are usually necessary when applying a high gauge or low density foam and this situation is representative of UK requirements. Consequently, wherever a gelled foam is used in the UK the filler level is less than 120 phr or includes natural latex, or both. In all cases silicone is added.

Processing is not too difficult provided certain key operatives, such as the plant foreman or foamer, have the necessary skills and know the limitations. For example, it is important to identify whether the ammonium acetate gel level is correct immediately after the infra-red gelation zone. To do so, they must recognise the symptoms of undergelation (ranging from large wide cracks through to an orange-peel surface with no surface cracks — both extremes are accompanied by severe shrinkage or foam collapse) and overgelation (visible as small surface cracks at 90° to the machine direction and often referred to as crazing). In either case corrective action must be taken swiftly to avoid downgrading expensive carpet. In this respect, laboratory tests on a compounds 'gel range' is an important tool to be used in setting the initial gel level, which for plant start-up should be slightly higher than the mid-point. Gel range, incidentally, is the operational range of ammonium acetate additions which produces a satisfactory product without either over- or undergelation. An example of a good gel range would be 3 to 5 per cent wet on wet, where anywhere inbetween would produce an acceptable product. A poor gel range may be only 3 to 3·5 per cent wet on wet, with the result that on machine the gel state may oscillate from 'over' to 'under' while perhaps never achieving consistent running between the two outer limits.

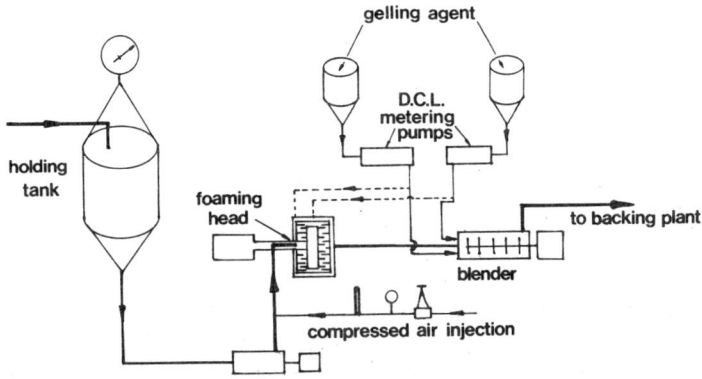

FIG. 5.4. Automatic foaming unit.

Mixing of the gelling agent is directly into the already aerated compound either by injecting into the foaming head or into a blender situated downstream of the foaming head (Fig. 5.4). All foaming and mixing is carried out under relatively high back pressure which can be varied by adjusting the length or bore of the delivery hose. High back pressure aids mixing of both air and gelling agent. Coating onto carpet is as previously described. Gelation is initiated in the infra-red zone and should be complete on exit. Drying and vulcanisation time varies according to a multitude of factors such as oven temperature, compound solids, foam density, applied weight and foam thickness. Typically though, a cure time of 12 min at 150°C would be expected for a 3 mm foam at 70 per cent solids and 700 g/m² applied weight.

The sodium silicofluoride gelation process

Sodium silicofluoride (NSF) functions as a delayed action gelling agent and is used only in the German, Austrian and Swiss carpet industries to any great extent. The reason, other than quality or perhaps historical precedent, is not clear because it is the most difficult to control of all the foam processes used by the carpet industry.

Gelation is brought about by the hydrolysis of NSF in water to liberate hydrofluoric acid (HF). In combination with zinc ions, the HF reacts with the stabilising fatty acid soap to cause a controlled coalescence at room temperature. But in the carpet industry there is insufficient time to allow a room temperature gelation and it is, therefore, treated like all other foams on carpet by passing it through the infra-red zone to accelerate gelation. This heat treatment, along with the presence of calcium carbonate filler, combine to produce the NSF's characteristic leathery skin, often termed 'elephant hide'. The mechanism for the formation of such an unusual skin is not clearly understood but it is certain that it is a consequence of HF reacting with calcium carbonate to liberate calcium ions, which in turn generate sufficient instability to cause rapid surface collapse when exposed to high intensity heat. Elephant hide is not formed if calcium carbonate filler is omitted or substituted by a clay.

TABLE 5.6
NSF gel foam recipe

	Per cent activity	Parts by mass (dry)
Styrene–butadiene latex	66	100
Potassium oleate	20	1·75
Sulphur	50	2·0
ZDEC	50	1·0
ZMBT	50	0·75
DPG	40	0·75
Antioxidant ?	40	0·75
Zinc oxide	50	3·0
Calcium carbonate filler	100	50
Polyacrylate thickener	16	to viscosity 3 Pa.s
Sodium silicofluoride (NSF)	25	3·5

Formulating for an NSF foam is relatively simple even if the subsequent compound processing is not. A typical recipe uses only an SBR latex (Table 5.6) designed specifically for this process. Standard SBR latices invariably do not work satisfactorily because they either do not form a uniform leathery skin or their gel range is too narrow. Soap at a low level is added for foamability and stability. The usual accelerators and other vital curatives are added along with a gel stabiliser, invariably DPG (diphenyl guanidine). The final additions are the calcium carbonate filler and viscosity modifier which can be methyl cellulose or polyacrylate.

There is no maturation of the compound and, in fact, it is used almost immediately whenever possible. Foaming is achieved in the usual way and the NSF dispersion is added to the foam via a blender. Processing is found to be easier at temperatures less than 20°C since this gives a wider gel tolerance. Often, therefore, carpet mills operating this process will chill the compound prior to foaming.

NSF compounds are expensive (entirely due to the low filler level) and are, therefore, not used indiscriminately. Generally it is applied to high quality carpets which have to pass the Rollstahl[9] test (where delamination failure must be avoided) or installations involving gluedown to the floor where NSF foam is supreme in its ability not to absorb adhesive.

The non-gel foam process

Thin sectioned foams were first produced by the non-gel foam process back in 1929[10] but it was not until the latter half of the 1960s that the principle became a commercial reality, following a patent by Crown Rubber.[11] Quite independently the technique had also been discovered by several other technologists, particularly in the UK, and the system was already starting to dominate the foam underlay and carpet markets by the time the patent became public knowledge. It would appear to be an unusual coincidence that so much activity should suddenly concentrate on the same technique but in fact the reason was connected with a revolutionary foam process developed by Dow.[12] In 1963 Dow put onto the carpet market a specially designed carboxylic SBR latex that could be used to manufacture thin sectioned foams. It was, therefore, ideally suited for carpet backing. The system offered many benefits but the most important were that no gelling agent or complicated sulphur cure paste was required. Instead the crosslinking was via the carboxyl sites on the polymer using a melamine formaldehyde (MF) resin. The MF resin was also accorded with the role of gelling agent.

Many latex technologists, whilst not disputing the role played by MF resin, were inclined to draw the conclusion that the system's excellent foam stability also played a major part. Therefore, on adding the same powerful surfactants, as used in the Dow system, to conventional soap-based SBR latices, and also natural latex, it was found that satisfactory foams could be made provided the gauge was restricted to not greater than 1·5 cm. No gelling agent had been added and in any case it would have had little effect owing to the presence of significant quantities of surfactant.

It is, therefore, obvious that the non-gel process functions in a completely different manner to gel systems. Coalescence is not brought about by destroying the coulombic repulsion forces which exist between latex particles. Instead, it relies upon increasing the stability of the water/air interface so that on raising the temperature to evaporate off the moisture, the foam cells remain stable allowing the solids to increase without foam collapse. At some point, before all the water is removed, the latex particles contained in the cell walls of the foam coalesce to form a continuous phase. Coalescence is much less uniform than with true gelation since the surface exposed to the heat source will coalesce and dry first whilst the part farthest from the heat will be last.

This gradation is readily visible in the end product where the cellular structure at the surface is relatively fine whilst at the carpet interface the structure is coarse. Gauge expansion is a further illustration of the gelling or coalescing mechanism. Conventional gel foams shrink on gelation resulting in a product some 10–20 per cent thinner than spread. Non-gels, on the other hand, expand by 10–20 per cent during the drying process.

In formulating for non-gel foams, it should be pointed out that only high solids SBR latices are illustrated in this section but it should be mentioned that natural latex can be processed by this technique and also that there are special carboxylic SBR latices designed exclusively for the manufacture of non-gel foams. Table 5.7 gives a typical formulation based on high solids SBR latex. The succinamate surfactant is the crucial ingredient which enables the non-gel system to function satisfactorily by building in the very high foam stability required. Also included is a lauryl sulphate surfactant which acts as a foam booster by increasing the foaming rate. The inclusion of a polyphosphate is most important because it functions as a sequestering agent as well as enhancing high temperature foam stability. Its omission can often lead to foam collapse in areas last to dry, i.e. the foam interfacing with the carpet. The cure system is typically sulphur and ZDEC-based with a relatively low level of thiazole. High levels of zinc oxide are not necessary, because zinc is not

TABLE 5.7
Non-gel foam recipe

	Per cent activity	Parts by mass (dry)
Styrene–butadiene latex	66	100·0
Disodium octyl sulphosuccinamate	35	4·5
Sodium lauryl sulphate	28	0·5
Sodium hexametaphosphate	25	0·5
Water	—	to 78 per cent TSC
Sulphur	50	2·0
ZDEC	50	1·0
ZMBT	50	0·75
Antioxidant[a]	40	1·0
Zinc oxide	50	2·0
Calcium carbonate filler	100	200·00
Polyacrylate thickener	16	to 5 Pa.s.

[a] Usually quinoline type for heat resistance.

required to take part in a gelling mechanism, and 2 phr or even slightly less is a common level. Like the other systems mentioned, the curatives, etc., would be supplied to a carpet mill in the form of a cure paste. The level of filler addition at 200 phr is much higher than gel foams and is one of the main reasons for the success of non-gel foams. Exactly why it should accept such large quantities of filler is due in part to the fact that the stress-inducing gelling step, which is apparent in the zinc ammine and NSF processes, is not involved. Also the non-gel cellular structure results in added strength which allows greater tolerance on filler dilution. To complete the compound, polyacrylate thickener is added to adjust the viscosity.

Non-gel foam technology has several fundamental advantages. First, its compound cost is low because of the influence of the high level of filler it can accept. Second, it processes much more simply because there is not a complex gelling process involved.

Easy processing also applies to low density foams and, therefore, allows even lower raw material costs because the applied weight per unit area can be reduced. Finally, as a result of the higher compounds solids, non-gel foams dry and cure faster than their gelled counterparts.

To complete the picture, it is worth mentioning that poor physical properties (due entirely to filler and density levels) and its water absorption characteristic (caused by the presence of high surfactant levels) represent the two worst features of non-gel foam technology.

Choice of foam system

All three foam systems are used to a greater or lesser extent in the major tufted carpet markets. The selection of which one to use is determined by three major factors — technical requirements, production simplicity or just good old fashioned minimum costs. Often, unfortunately, technical requirements have taken a back seat with the inevitable result (and hindsight) that consumers have become aware and are alert to a falling in general quality standards. Even high quality foams are affected by the consumers distrust of foam backed carpet and consequently some manufacturers are showing increasing interest in mechanically foamed polyurethanes as a replacement for latex foam. Needless to say, this situation is recognised and action by latex manufacturers is already in hand.

In an attempt to quantify foam quality, the physical properties of the three systems are compared in Table 5.8 at the same applied weight and density of 700 g/m^2 and 0.23 g/cm^3, respectively. To ensure a balanced view is obtained by the reader, filler level and raw material cost ratios are also quoted. The first point of note is that tensile strength is significantly higher for the lower filled NSF foam, followed by the zinc ammine gelled foam with the non-gel bringing up the rear. The same trend is true for delamination strength and surface abrasion resistance. Elongation-at-break, however, does not follow the same trend because zinc ammine gelled foam is only as good as the non-gel, the reason being due to minute cell rupture typical of the zinc ammine process. Compression resistance is not widely different between any of the systems, but the figures do indicate that, contrary to most peoples expectations, foam

TABLE 5.8

Physical property comparison of non-gel, zinc ammine and NSF foams at a density of 0.23 g/cm^3 and 700 g/m^2 application weight

	Non-gel	Zinc ammine	NSF
Filler level (phr)	200	120	50
Tensile strength (kPa)	42	77	190
Elongation-at-break (per cent)	160	150	290
25 per cent compression modulus (kPa)	28	30	35
Delamination (kg/5 cm)	0.7	1.30	2.3
Surface abrasion resistance (number of strokes)	4.0	14.00	30
Water resistance	Poor	Good	Excellent
Raw material cost ratio	1.0	1.33	1.90

hardness is hardly affected by increasing filler of the type generally used in the carpet industry. Fine particle fillers obviously behave differently. Water resistance, an important property for carpets glued directly to the floor, is considerably superior for gel foams, particularly NSF.

Clearly non-gel foams perform badly in this comparison but it should be emphasised that this is not a feature of their processing technique. The real reason lies in their filler tolerance which inevitably dilutes the physical properties of the starting polymer. At equivalent filler level their dynamic properties can be better than zinc ammine and as good as NSF. They will still have poor water absorption, however.

Finally, in summarising these features it can be said that in todays market whenever technical performance is the major criterion (i.e. heavy duty contract) then invariably a gelled foam is used. Non-gels at low filler levels are used only for contract installations where water (or adhesive) absorption is not a problem. The domestic market is served by all systems, but this is where the major advantage of non-gel foam technology comes into its own, particularly for carpet qualities at the low to middle end of the market.

REFERENCES

1. *Intercontuft Carpet Statistics for 1978.*
2. ASTM D2859-70, *Methanamine Pill Test.*
3. BS 476 Part 1, *Surface Spread of Flame.*
4. BS 4790, *Hot Metal Nut Test*, 1972.
5. *High Solids Latex Compounds for Jute Lamination*, Product Information Sheet, Southern Latex, Georgia, USA.
6. CANT, S. B., *Rubb. Dev.*, 1959, **12**, 2.
7. BLACKLEY, D. C., *High Polymer Latices*, Vols. 1 and 2, Applied Science Publishers Ltd, London, 1966.
8. British Patent Specification 1,239,554.
9. Draft DIN Specification 54324, *Testing and Evaluation of Textile Floorcoverings for Castor Chairs.*
10. US Patent Specification 1,777,945.
11. British Patent Specification 1,105,538.
12. US Patent Specification 3,215,647.

Chapter 6

SYNTHETIC LATEX BINDERS

R. CHARLTON,† J. T. BOWDEN and B. SMITH
Revertex Ltd, Harlow, UK

In this chapter a wide range of latex applications and very diverse technologies are considered. Both in the Paper and Textile Industries, synthetic latices are used extensively as binders, whether it is primarily for clays in the surface coating of paper, for asbestos fibre in production of gasketing base or for bonding during manufacture of bonded fibre fabrics such as wiper cloths, etc. In each case the latex is expected to satisfy a number of requirements in addition to acting as a binder, and these dictate the design and choice of the most suitable product.

Five key market areas are discussed. These are:

pigment coating of paper and board,
latex 'wet-end' addition,
paper saturation,
bonded fibre fabrics, and
woven fabrics.

SYNTHETIC LATEX IN PAPER AND BOARD COATING

Today a whole range of polymers are used in paper coating, but in Europe, considered the most technically demanding market, carboxylic–styrene–butadiene latices find the widest application. Before examining the role of polymer latices in this field the questions of why and how paper is coated must be answered.

There is a variety of functional coatings which can be applied but the most common is aqueous pigment coating. This provides a smooth

† Present address: STC Ltd, Temple Fields, Harlow, Essex CM20 2AH, UK.

surface with controlled ink receptivity to enable the complete transfer of half-tones during printing. The coatings consist of pigments,[1] binders, and water, together with minor quantities of functional ingredients such as defoamers, flow modifiers, and fluorescent whitening agents.

Coating mixes are normally prepared in batches of several tonnes using powerful and sophisticated agitators to deflocculate the pigments to give a smooth dispersion. The major pigment is finely ground china clay (kaolin),[1] used either alone or in conjunction with other materials chosen to impart specific properties to the coating surfaces, e.g. Satin White (calcium sulpho-aluminate) and titanium dioxide are used to increase gloss and whiteness, whilst calcium carbonate can impart a matt finish.

The function of the binder is to fix the pigment particles to the paper surface and each other in a manner best visualised as spot welding. However, the nature and level of binder present will influence coating mix rheology and modify surface printability characteristics.

The coating process

The coating process can be considered in three stages namely application, metering and levelling. Of the many coater types,[2] some deal with each process stage separately whilst others complete the stages almost simultaneously. Perhaps the most commonly used coater is the inverted trailing blade. This can be installed in line with the paper making machine ('on-machine') or as a separate piece of equipment ('off-machine') perhaps serving two paper machines. There are several commercial designs of this coater operating on the same principle. An applicator roll running in a bath of coating mix transfers the mix to the paper web. In close proximity, a metal blade meters off excess mix into a recirculation system. The weight of coating applied is controlled by the blade angle and pressure, together with coating mix solids.

A recent development of this system uses a longer more flexible blade, controlled near its tip by a pneumatic tube. Much lower blade angles are used and considerably higher coat weights are achievable (e.g. 0·025 kg/m^2 or higher).

Typically, the trailing blade coater is used to produce large tonnages of paper for magazines, catalogues and labels for cans, bottles, etc. In contrast, the metering bar coater, for example, is normally found on-machine, and is principally used for coating board.

Excess mix is applied by an applicator roll and subsequently metered by a bar rotating against the direction of travel of the web. Coat weight is controlled by mix solids and web tension. A typical product of this type of coater would be a decorative carton or display card.

There are many other coater designs each of which has its own specific needs in terms of coating mix rheology and solids. These needs, in turn, dictate the choice of binder and cobinder and their ratio. For example, an 'all latex' binder system may be suitable for an off-machine blade coater, whilst a protein/latex blend could be preferred for an on-machine metering bar. Table 6.1 shows typical viscosity and solids data for various coaters.

After coating and drying the paper is usually calendered[3] to further improve its smoothness. If a full gloss finish is required super calendering is performed. Here the coated surface is mechanically polished between rolls under pressure and often at elevated temperature. This process realigns the pigment particles to give a smooth glossy finish.

The printing process

In order to fully understand the demands made on paper coating latices it is necessary to consider the salient features of the three common printing processes.[4] These processes are characterised by the method in which the image to be produced is defined on the printing plate. In letterpress, the image areas are raised above the non-image areas, in the same way that characters are defined on a type-writer. A lithography plate is flat, with the image and non-image areas defined chemically.

TABLE 6.1
The relationship between coating machine and physical properties[a]

Coating machine	Typical viscosity[b] (mPa.s)	Typical solids content	Maximum coat wt (g/m²/side)	Maximum speed (m/min)	On/off machine
Brush	150	40	30	100	Off
Air knife	150	40	30	350	Both
Transfer roll	20 000	60	16	300	On
Size press	150	15	5	300	On
Meter bar	400	35	8	250	Both
Billblade	300	56	10	200	On
Trailing blade	3 000	58	10	1 000	Both
Extended blade	1 000	58	40	1 000	Off

[a] Reproduced from *Paper*, Vol. 186, No, 9, 1976.
[b] Figures quoted are Brookfield viscosities.

In gravure the image takes the form of a series of dots engraved into the plate surface. Each printing process requires a different type of ink. Letterpress inks are viscous and tacky, drying by oxidation. Lithography uses similar inks to letterpress but with higher viscosities and tack rating. Gravure inks are generally solvent-based and have very low viscosity (~ 100 mPa.s) to enable transfer from the etched cells of the plate to the paper surface.

Each process involves splitting an ink film, with part transferring to the paper and part remaining on the printing plate. As the film splitting occurs a rupturing force is generated at the coating surface. If the bond between the coating layer and the base paper is weaker than the force, splitting will occur at the interface rather than within the ink film. Coating will be torn from the surface, adhering to the printing plate and leaving craters with paper fibres exposed at their base. This phenomenon is known as 'dry picking'.[5]

Because of the low viscosity of gravure inks the film splitting force normally is too low for dry picking to occur.

In the lithography process the plate is first damped with a water-alcohol mixture which adheres to the chemically defined non-image areas. Oil-based ink is then applied and only adheres to the image since it will not mix with water. The paper surface is then printed with both water and ink simultaneously.

Illustrations that contain two or more colours are achieved by overprinting combinations of images. This is achieved by having a series of printing stations in line. An area to be printed only with the last colour in a sequence will therefore have been wetted as a non-image area several times before ink is applied to it. When this happens the coating is subject to the same rupturing forces as in letterpress but when thoroughly wetted. A second type of coating failure can now occur normally; this is called 'wet picking'.

In gravure printing, the degree of contact between the ink in the image cells and the paper surface is of prime importance since this will control the degree of ink transfer.

The main requirements of the coated surface for the three types of printing are therefore:

Process	Requirement
Letterpress	Dry pick resistance, smoothness, absorbency
Lithography	Wet and dry pick resistance, absorbency
Gravure	Smoothness

Coating mix formulation

Adjustment of the coating mix formulation will fulfil some of the above requirements but the chemical nature of the binder used is of prime importance. Choice of binder type is a compromise between the rheological needs of the coating process, the blend of printability properties required for the finished surface, and economy. In commercial practice polymer latices are used in conjunction with other co-binders, both natural and synthetic.

Letterpress and lithography require higher pick resistance than gravure and so a higher binder level or higher power binder must be used. Table 6.2 gives an indication of binder levels in typical basic recipes.

Comparison of latex with other binders

Latices, as supplied, are low viscosity liquids and can easily be handled on automatic mix preparation systems. Starch, casein and carboxymethyl cellulose, on the other hand, all require some form of solution preparation which can involve the use of specialised equipment.

Latex has a higher binding power than the natural binders and can, therefore, be used at a lower level in the mix formulation resulting in improved optical properties.

Apart from the self-thickening types, latices have little or no effect on mix viscosity. This enables the use of higher solids mixes with a consequent saving in drier energy requirements. Improved printability performance may also result.

Latex is less susceptible to bacterial or fungal attack than natural binders. It tends to show improved flexibility, glueability and varnishability in coated carton board.

Latices can be designed to give a wide range of printability perfor-

TABLE 6.2
Relationship between binder levels and printing process

Binder/process	Gravure		Letterpress				Lithography		
Latex	7	6	8	11	7	9	14	12	12
Carboxymethyl cellulose	—	—	—	2	—	—	—	2	—
Starch	—	4	6	—	—	7	—	—	—
Casein	—	—	—	—	5	—	—	—	6
Emulsion thickener	0·3	—	—	—	—	—	0·3	—	—

Numbers refers to per cent of binder on pigment.

mance. In contrast, conventional latices have low water retention in comparison with starch and protein. Thus, in practice, to obtain the required rheological and water retention characteristics, latices are frequently used in conjunction with natural co-binders.

Latex types

Within Europe carboxylic–styrene–butadiene latices are the most widely used, followed by carboxylic–acrylic copolymers. In comparison polyvinyl acetate and vinyl acetate–ethylene copolymers account for a significant percentage of consumption in the USA. This interesting divergence of use is generally explained by the different quality and performance standards required in the two markets.

The continued growth of carboxylic–styrene–butadiene latices is mainly due to their cost efficiency. Generally they have high binding power and impart good water resistance to the coating. XSBR coatings tend to be susceptible to heat and light ageing and historically have been considered more critical in odour for food packaging. Recently a number of very low odour XSBR latices have been made available and are finding use in food packagings.

Carboxylic–acrylic latices are typically copolymers of acrylic esters with styrene. They find use where freedom from light ageing and low odour are of prime importance. Generally they tend to give higher gloss and ink absorbency than XSBR but with lower binding power.

Polyvinyl acetate latices are noted for the open structure they confer to the coated surface. This is claimed to be an advantage in papers for heat set web offset printing[6] and in conferring good glueability to carton boards, particularly where acetate-based adhesives are employed. However, they have relatively low binding power and water resistance.

Polyvinyl alcohol in its fully hydrolysed form has very high binding power and imparts excellent brightness. However, difficulties in mix preparation (pigment shock)[7] and poor high shear rheology at high solids have tended to limit its use to a cobinder in speciality papers.

Polyvinylidene chloride latices are widely used in wallpaper top coatings where exceptional water resistance is needed.

'Sole latex' binders are comprised of a latex binder with the thickener 'built in', usually in the form of a high level of carboxylation. The products which can either be styrene–butadiene- or acrylic-based are activated by the addition of alkali, raising viscosity. Such systems have been criticised as inflexible since changing the latex level in the mix also changes viscosity. This problem may be overcome by using low levels of highly active thickeners in conjunction with normal latex binders. These thickeners are typically very highly carboxylated acrylic copolymers.[8]

POLYMER COMPOSITION

The combination of properties shown by a polymer is derived from the mixture of monomers used in its production. The effect on printability performance of the use of the common monomers is illustrated below.

Monomer	Property imparted
Styrene	Stiffness, ink absorbency, gloss
Butadiene	Binding strength, flexibility, fold endurance
Acrylonitrile	Solvent resistance, ink absorbency, stiffness
Acrylic esters	Resistance to light ageing
Unsaturated carboxylic acid	Binding power

Most paper coating latices contain at least two main monomers together with lower levels of other monomers to modify performance. The ratio of the two main monomers will have a large effect on the polymer printability performance.

EFFECT OF MONOMER RATIO IN STYRENE-BUTADIENE LATICES

Polystyrene is far too stiff and brittle to use as a pigment binder while, conversely, polybutadiene is too soft. In combination, however, a wide range of performance can be achieved.

For maximum binding power, or dry pick resistance, an optimum styrene level of around 50 per cent exists. Above this level the polymer loses its extensibility and hence its ability to resist the force generated by ink film splitting in the printing process.

• Increasing the styrene level above 50 per cent significantly raises ink absorbency. It is argued that this is due to the decreasing tendency of the polymer to film formation, producing a more open and porous surface.

Thermoplasticity rises with increasing styrene level. In the super calendering process, under the influence of temperature and pressure, this thermoplasticity permits the realignment of the pigment particles to produce a higher gloss.

Within the range of monomer ratios commonly used in pigment coating latices, wet pick strength is little affected.

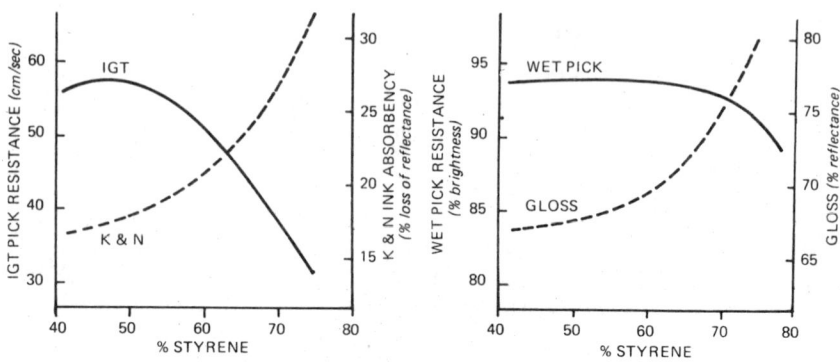

Fig. 6.1. Effect of styrene content on coated paper properties.

The effect on printability of styrene levels between 50 and 70 per cent in a styrene–butadiene latex is illustrated in Fig. 6.1.[9]

With the use of a third monomer in the copolymer a further range of properties can be achieved. Acrylonitrile, for example, has a T_g similar to styrene but contributes to both binding power and ink absorbency. If it is used as a partial substitute for styrene both dry pick strength and ink absorbency can be raised. These two properties are directly opposed in simple styrene–butadiene copolymers.

Styrene–butadiene copolymers tend towards block rather than random structure. This is due to their reactivity ratios. Polybutadiene is still unsaturated and, therefore, liable to oxidation, which explains the relatively poor heat and light ageing of SBR. By comparison butadiene–methyl methacrylate and styrene–butyl acrylate copolymers are almost entirely random and give good ageing characteristics.

Unsaturated carboxylic acids are arguably the most important monomers in paper coating styrene–butadiene polymers. Whilst typically present at less than 5 per cent of the total monomer they have profound effects on both rheology[10] and printability performance. Carboxylic acids increase the copolymer T_g and hence raise the stiffness. They improve the adhesion to cellulose fibres and to hydroxyl sites on the pigment and they provide sites for crosslinking the polymer film to improve water resistance.

Both chemical and mechanical stability of the latex are improved by carboxylation, each property being dependent on both acid level and type. A high acid level (greater than 5 per cent of total monomer) will impart alkali sensitivity to the polymer, giving a significant rise in mix viscosity above a pH of 7·0. Water retention, a property closely allied with machine runability, is also dependent on carboxylation..

Other polymer variables influencing coating performance

For any given monomer ratio there is an optimum molecular weight band for both wet and dry pick resistance. The choice and level of chain transfer agent is therefore important. Certain chain transfer agents can produce odoriferous side products in the polymerisation, which would be detrimental to the residual odour of the finished paper coating.

Surfactant type and level will influence the latex stability, water sensitivity, and foaming characteristics. Additionally, during polymerisation it will control the polymer particle size and size distribution. A small decrease in particle size gives a large increase in the number of polymer particles in each unit weight of product. As particle size decreases both wet and dry pick resistance increase. However, to a lesser extent gloss and ink absorbency decrease as the pigment particles become more closely and tightly bound together and so more difficult to realign during calendering. Typical particle sizes of commercial products vary from 140 nm to 230 nm.

In addition to the effects of polymer chemistry on coating performance it must be realised that the physical conditions of polymerisation are also important. Reactor and stirrer design, reaction rate and temperature can all influence polymer properties.

Commercial products

With such a wide range of printability performance available from, for example, carboxylic–styrene–butadiene copolymers, an equally wide range of products is available from the manufacturers. Most manufacturers offer two products for the mainstream applications together with a number of specialised grades. In the case of styrene–butadiene latices, these two products can be loosely identified as:

Product 1 — Higher butadiene, very high binding power, low ink absorbency, medium gloss.

Product 2 — Higher styrene, medium dry pick resistance, high wet pick resistance, high gloss, high ink absorbency, good stiffness.

In recent years more and more emphasis has been focused on the toxicological effect of chemicals. This together with changing end-use requirements (for example, coated board trays to contain food whilst it is cooked in a microwave oven), have placed many restrictions on the choice of polymer ingredients. Latex manufacturers have responded to these demands and can now offer latices with a full range of printability performance as well as compliance with the various toxicology regulations.

USE OF LATEX IN PRODUCTION OF PAPER AND RELATED MATERIALS.

The incorporation of latex in paper, as opposed to its application on the surface only, can dramatically improve its properties. Tensile strength, edge-tear resistance, water and oil resistance and flexibility are among the properties which may be enhanced. For certain fibres, for example asbestos, latex may be necessary in order to bind the fibres together and not just to improve the sheet properties.

Latex types

Many types of polymer latex are used in paper, ranging from soft elastomers to hard thermoplastic polymers. The most common of these are shown in Table 6.3, together with the key paper properties which they may influence.

TABLE 6.3
Latex type and influence on paper property

Latex \ Paper Property	Softness	Medium hardness	Hardness	Flexibility	Elasticity	Toughness	Fold endurance	Tear resistance	Water resistance	Oil-solvent resistance	Resistance to light ageing	Wet strength	Rigidity	Tensile strength	Thermoplasticity
Natural rubber	×			×	×		×								
Polybutadiene	×			×	×		×				•				
Low-styrene styrene–butadiene copolymers				×	×	×	×	×							
Acrylic copolymers	×	×		×			×			×	×	×			
Vinyl acetate–acrylic copolymers				×						×					
Acrylonitrile copolymers				×					×	×		×		×	
Polyvinyl acetate			×												×
High-styrene styrene–butadiene copolymers			×										×	×	×
Polychloroprene									×	×	×				•

Method of latex addition

Latex is added to the sheet in two basic ways: (a) during the sheet making process, known as 'wet-end addition'; or (b) after the sheet has been formed, generally by a saturation procedure.

WET-END ADDITION

This is the process of adding latex to a fibre slurry or 'stock' at some stage prior to its application to the wire on which the sheet is formed.

In a typical paper making process, the fibre passes through several preparation stages. The initial pulping stage reduces the fibrous material, for example, wood, to fibres. The fibres are then dispersed in water. The length and degree of fibrillation of the fibres are modified by beating in which the fibres are subjected to cutting and bruising by dull blades. Finally the stock is cleaned and diluted before being passed to the wire where the water drains from the fibre to form the sheet. The foregoing stages constitute the 'wet-end' of the paper making process, that is those stages in which the fibre is present as a suspension in water. Details of the processes are given in an authoritative treatise.[11]

Addition of latex can take place at any convenient point during stock preparation, provided there is sufficient agitation. The beater, the fan-pump and the headbox may be used, the former giving rise to the term 'beater-addition'. However, deposited latex can be dislodged by shear force and hence it is desirable to add the latex at a point as close to the wire as possible.

The action of the latex is to deposit on to the fibre and thus to partially replace some of the fibre–fibre bonds (which would have been present in the sheet), by fibre–latex bonds. This leads to increased tensile strength and sheet properties related to those of the latex being used. In order to deposit readily on the fibre, latices employed in wet-end addition need to be relatively unstable — hence, carboxylic latices are usually unsuitable. Where a carboxylic type is used, the time taken for the latex to deposit is generally longer than for a similar uncarboxylated latex.

There are a number of methods of adding the latex to the fibre, the precise technique chosen being dependent on the type of fibre — cellulose, asbestos, leather and man-made fibres being the commonest. These may have a natural charge, positive or negative, or have no charge at all. Since the majority of commercial latices are anionic (negatively charged), it follows that deposition will only take place readily on to positively charged fibres. Fibres which are negatively charged or uncharged require special agents to effect deposition. The two main agents which are used are Papermaker's alum · (aluminium sulphate) and cationic resin deposition aids.

Papermaker's alum. The most widely used procedure utilising Papermaker's alum is known as the direct alum process. The action of alum is complex but is thought to involve both an interaction of the alum with the latex emulsification system, causing destabilisation of the latex, and also neutralisation of charge of the fibre.

In a typical procedure, the stock pH is first adjusted to be compatible with that of the latex (e.g. pH 8–9). The latex, diluted to 20 per cent solids, is added with gentle agitation. This is followed by a 5 per cent alum solution until complete deposition has occurred, shown by the clarification of the backwater. In this and any other deposition process, excessive shear force on the stock after the deposition should be avoided. Use of latex as an additive results in slower drainage of the stock on the wire, compared with unmodified stock. This leads to lower machine speeds.

Cationic deposition aids. A number of polymeric materials fall into this category, their modes of action being similar. They are deposited on to the fibre to which they impart a positive charge. The anionic latex then deposits on the fibre without difficulty.[12]

The processes which are based on these materials include the Bardac Process,[13] which uses a melamine–formaldehyde resin, and the Snyder Process based on a phenolic resin.

An illustration of the technique is given by the following method based on the Bardac Process. The pH of the stock is adjusted to approximately 5 with Papermaker's alum, to be compatible with the deposition aid. The cationic resin is added with gentle agitation and allowed to deposit for a suitable period, e.g. 30 min. The pH is re-adjusted to 8 and the diluted latex added, again with agitation. Complete deposition of the latex is indicated by a clear backwater.

Cationic deposition aids are particularly useful for uncharged fibres, such as man-made fibres, and those with a low surface charge. Also, these techniques allow a greater quantity of latex to be added to the fibre and the deposited latex is more resistant to shear forces.

Fibres and Applications

The types of fibre used in wet-end addition processes and the products made may now be discussed in greater detail.

Cellulose. Although cellulose fibre forms paper sheets of great strength without addition of latex,[14] its inclusion can confer useful properties, many of which have been mentioned previously. Latex is generally used to manufacture speciality papers such as gaskets, carbon and filter papers. The wide variety of cellulose materials available to the papermaker — woodpulp, cotton, recycled waste, etc. — are used to produce a wide variety of papers, the exact choice of fibre being determined by the product. The amount of latex which may be added is up to 30 per cent of the weight of the fibre on a dry basis.

Cellulose has a natural negative charge, although the degree of charge is variable — for instance, long fibres possess a low charge — and the charge is not necessarily evenly distributed along the fibre. Fibres with a high degree of fibrillation ('split-ends') have a greater charge density associated with the fibrils. For papermaking purposes the amount of fibrillation of the fibres is generally increased, by the beating process, in order to improve sheet formation. Thus, addition of latex is facilitated.

For many purposes, the direct alum process is suitable for the deposition of the latex, although for fibres of low charge a cationic deposition aid is necessary.

Cellulose-based gasketing materials which require good tensile strength and, particularly, good resistance to oils and water at elevated temperatures tend to contain a medium–high acrylonitrile–butadiene copolymer at about 15 per cent dry polymer on fibre. Natural rubber or chloroprene on the other hand are preferred for shoe insole production, for example, where good flexibility and resistance to dampness are important. At the other extreme thermoplastic high styrene SB copolymers are chosen, to give the balance between stiffness at room temperature and mouldability when subjected to heat (~ 100–$105°C$) and pressure (~ 300 Pa), in the manufacture of 'heat-mouldable' boards. The latter are used in suitcases and for various items in the interior trim of motor vehicles; low grade waste cellulose helps to yield a relatively low-cost product.

Many types of paper are produced for special applications using wet-end addition processes. Often quite low levels of latex (5–10 per cent on the fibre) are used. For example, disposable towels, filter papers, and floor tile base may be made using a beater addition process.

Asbestos. Unlike cellulose, asbestos fibres will not form strong sheets without a binder. Latex is widely used for this purpose. There are several forms of asbestos fibre, but the ideal one for beater addition is Chrysotile, because of its strength, flexibility and natural positive charge which allows anionic latices to deposit on to it without the need for a deposition aid.

Beater addition of latex to an asbestos stock is a relatively simple process. The stock is dispersed and the latex is added to the stock with gentle agitation. There are, however, a number of factors which need to be controlled in order to obtain a finished sheet of the desired strength, and at the same time obtain machine speeds which are sufficiently fast. Latex causes the asbestos fibres to agglomerate; the degree of agglomeration determines the drainage rate of the stock on the wire. Highly agglomerated stocks drain quickly but give poor sheet formation whilst less agglomerated stocks give better sheet formation but slower machine speeds. The drainage rate of the stock is therefore of prime importance, and is affected by, among other things, the type and amount of latex, the type of fibre, speed of agitation and chemical additives present.[15]

The three principal areas of use of latex wet-end addition to asbestos are in production of gasketing base, flooring substrate and roofing felts. In gasketing similar latices may be used as with cellulose. For asbestos flooring substrate a medium styrene–butadiene copolymer at 15–20 parts on fibre can provide the necessary toughness with low elongation, good tensile strength and resistance to water up-take. In roofing felts, where water resistance is also important (coupled with good weathering performance and flexibility), a soft styrene–butadiene copolymer at 15–25 parts on fibre is suitable.

Synthetic fibres. Some non-woven fabrics are made by a beater addition process from nylon or rayon, with a proportion of cellulose. The cellulose is in the form of long fibres to strengthen the sheet. Non-wovens made in this way are termed 'wet-laid'. Dry-laid non-woven fabrics are dealt with elsewhere in this chapter. The higher the proportion of synthetic fibre in the sheet, the more fabric-like is its character.

A cationic deposition aid is used to deposit the latex on to the fibre because neither the synthetic fibres, being uncharged, nor the cellulose fibres, because of their length, are particularly receptive to latex deposition. Dilute stocks are utilised to prevent knotting of the long fibres.

Non-woven fabrics are formed on machines similar to those on which paper is made.[16,17] However, because of the dilution of the stock, the machine speed is slower to allow for the drainage of the greater amount of water. To enable a more concentrated stock to be used, a special process is required. In the Radfoam process,[18] the fibres are applied to the wire in a foam. The pseudoplastic properties of the foam ensure adequate dispersion of the fibres under high-shear stirring, whilst keeping the fibres separated during transportation to the wire.

Non-woven fabrics are generally used in disposable applications — clothing, disposable sheets, etc. Thus, the important requirements of the sheet are softness, flexibility, cheapness and, possibly, washability. Soft styrene–butadiene copolymers give flexibility and drape characteristics, whilst polyvinyl acetate is used where cost is the main consideration. A soft acrylic polymer is suitable where some degree of washability is required.

Reconstituted leatherboard made from scrap leather also requires latex addition by the wet-end process. This application is included in Chapter 11.

SATURATION

This is the process whereby latex is added to the paper web after the web has been formed. The web is immersed in a bath of latex, passed through nip-rollers or between knife-scrapers, to remove excess latex, and finally dried. The web may be saturated either when it is in the dry or in the wet state as it comes off the paper-machine, hence the terms 'dry-web' and 'wet-web' saturation. Wet-web saturation has the advantage that the web needs to be dried only once and hence is more economical, since the drying stage after the formation of the paper is omitted. However, there are two disadvantages of wet-web saturation. The web may not have sufficient strength to pass through the nip when wet and less latex can be incorporated in the sheet since a wet-web is less absorbent than a dry one.

The paper used for saturation should ideally be highly absorbent, of high bulk and possess sufficient wet strength. It is usually composed of cellulose-kraft, sulphite and rag being the common types. A proportion of synthetic fibres improves the absorbency.

The plant for the saturation process basically consists of a bath of latex through which the paper web passes. There are two designs of bath — horizontal and vertical.[19] The choice of bath is dependent on the length of time the web requires to be held in the bath, which is determined by the absorbency of the web. The horizontal type is used for short immersion periods.

The amount of latex which may be incorporated in the paper is very high — up to 100 per cent of the fibre weight. The latex pick-up is controlled by the solids content of the latex bath and the pressure on the nip. The latex solids content is relatively easily adjusted by the addition of water, preferably soft, and is the preferred method of control since excessive nip-pressures can cause sheet consolidation. Very dilute latex and high drying rates should be avoided, however, since this leads to latex migration. This occurs when the water in the wet sheet rises to the surface too rapidly, because the drying rate is too great for the amount of water contained in the web. The latex is carried to the surface by the water, leading to a deficiency in the centre of the sheet. Consequently, the sheet has poor internal strength and tends to delaminate. It also tends to stick to the surface of the drying cylinders.

The latex is not bound to the fibre as closely as in material produced by wet-end addition, since it does not interrupt the fibre–fibre bonds, and it has a greater degree of continuity.[11] Thus, saturated sheets are generally softer and more flexible than those produced by wet-end addition, and the sheet retains the strength of the fibre–fibre bonds. At very high levels of polymer, the sheet properties become more like those of the polymer and less like those of the original paper.

Latices and Applications

A latex for saturation should have a particle size of less than 300 nm to prevent blocking of the paper pores, and have good mechanical stability to prevent build-up on the rollers. An amount of surfactant in the latex and low latex viscosity will assist penetration. For high pick-up values, a high solids latex is desirable, if the viscosity requirements can be satisfied.

A wide variety of paper properties may be improved by latex saturation, as outlined in Table 6.3. Soft polymers, such as natural rubber, polybutadiene and soft SBRs are used where good stretch, flexibility and fold endurance are required. Such characteristics are useful in map-papers and synthetic leather-cloth which is in fact a latex saturated cellulose web coated with nitro-cellulose. Copolymers containing acrylonitrile are used for gaskets and masking-tape where oil and solvent resistance are required. These polymers also give good tensile strength and are particularly suitable for saturation due to their small particle size. For improved wet strength without the need for extra additives used in the production of, for example, abrasive base papers, a carboxylic–acrylonitrile copolymer is selected. Soft to medium-soft acrylic copolymers give good water resistance and edge-tear characteristics, the softer types giving better edge-tear properties at the expense of tensile strength. These polymers are used for strippable wallpapers and disposable medical products.

Comparison of wet-end addition and saturation

The choice of the process for the incorporation of the latex depends upon the sheet properties required and the equipment available.

Wet-end addition generally gives a tougher sheet because of the extensive fibre–polymer bonding present. Saturated paper on the other hand can give a stronger sheet because the fibre–fibre bonds remain intact and are reinforced by the polymer.

A great amount of polymer may be incorporated by saturation — up to 100 per cent on the weight of the paper compared with 50 per cent for wet-end addition.

Wet-end addition requires no special equipment other than that normally used for papermaking, making it a cheaper and simpler process than saturation. It is also a faster process, since the drying section of a standard paper machine has a greater capacity than that of a saturation plant. Further, wider webs can be produced. Wet-end addition is thus the more versatile process.

BONDED FIBRE FABRICS

The term 'bonded fibre fabrics' is explicit and is defined by the Textile Institute as 'a structure consisting of one or more webs or masses of fibres held together with a bonding material'.

The means of producing these fabrics include printbonding, impregnation, spray bonding, stitching and spun bonding. These various techniques comprise the 'dry-laid' systems of bonded fabric production (reference has been made to the 'wet-laid' process earlier in the chapter), and the means by which the fibres are prepared for final bonding is by fibre carding or air laying. In the former method the fibres are deposited parallel to the machine direction (although cross layering can be achieved if required), while in the latter case the fibres are distributed at random throughout the web.

Of the above mentioned techniques printbonding, impregnation and spraying all require a separate bonding agent in order to produce a serviceable fabric, while the fabrics produced by stitching or spun bonding processes do not.

Bonded fibre fabrics are used in a multiplicity of applications, e.g. wiper cloths, sanitary products, filter cloths, disposable household fabrics, shoe components, garment interlinings and many more. The production processes which employ polymers may be outlined as follows.

Print bonding
In this process the fibres are carded to produce either a parallel or cross-laid web. It is normal to use several cards aligned in tandem to build-up a web of the required weight. This web is then passed to a pre-bond or wetting out applicator roller and then onto the print bonding rollers, where the binder is transferred from a bath via an engraved pick-up roller onto the fabric which is nipped between the pick-up roller and a rubber covered pressure roller. The purpose of the pre-bond is to wet out the web in order to ensure good binder penetration.

The print bond process is used where the fabric must retain a soft hand, and as the whole fabric surface does not contain binder, a 'textile' feel is obtained due to protruding fibres.

The principal properties required from a binder for print bonding include:

high wet strength but rapid wetting out of dry film,
a soft film with good clarity and heat sealability,
low aqueous extractables and low toxicity, and
fast cure and high solvent and oil resistance.

With the polymer latices available today all the properties that are listed can be achieved but possibly cannot be optimised within one polymer system, at the most economical price. Thus, vinyl acetate–acrylic copolymers conferring heat sealability are used in sanitary fabric applications, and crosslinking styrene–butadiene rubbers and acrylic copolymers are preferred where solvent resistance is required, for example, in wiper cloths.

A typical formulation for use in print bonding would be:

	Parts wet	Parts dry
Water	295·8	—
Binder (SBR latex)	212·8	100
Antifoam	0·1	—
Crosslinking resin (melamine–formaldehyde)	2·2	2·0
Polyacrylate thickener	to required viscosity	
Total solids content (TSC)	19 per cent	
pH	8–9	

Binder application weights are in the region of 20 per cent of fibre web weights. The polymers used in print bonding are soft and the T_g ranges from $-60°C$ for the SBR types to $-24°C$ for the vinyl acetate acrylates.

Impregnation

The total impregnation of fibre webs is used where high application weights of binder are required, and the webs themselves contain a higher fibre weight.

The preparation of the webs, usually referred to as 'batts', takes one of two forms; either (a) loose laid fibres which are carded or random laid, or (b) needlepunch.

The method of impregnation is determined by the batt form, and is achieved either by a system of stainless steel wire belts or a standard total impregnation roller arrangement.

Batts that are not needled must be consolidated during the impregnation processes or they would disintegrate during immersion and expression. Consolidation is achieved by carrying the batt through the impregnation trough between two woven steel wire belts as illustrated in Fig. 6.2.

When needled batts are used, then impregnation can be achieved by a conventional expression roller system, which may be similar to Fig. 6.2 but without the wire belts.

Bonded fibre fabrics produced by total impregnation include:

(i) Carpet underlay — here the binders used are principally carboxylic SBR types of approximately 40–45 per cent styrene content. The binder is required essentially to impart high resiliency to the felt and thus create good recovery from static[20] and dynamic load.[21]

(ii) Needlepunch carpet — the binders in needlepunch impregnation have a similar monomer composition to those used in underlay. However, in this sector acrylic binders also find a limited use, e.g. in the 'W' processes.[22]

A combination of several properties is demanded for needlepunch carpets. This includes high abrasion resistance for the wear surface, good static and dynamic load recovery, good soiling resistance, low flammability and 'non-chalking', i.e. the absence of breakdown of the polymer film when abraded. This is of particular importance with polypropylene fibres.

(iii) Interlinings — acrylonitrile–butadiene copolymers of medium acrylonitrile content and acrylic copolymers are used to effect good water and solvent resistance and to impart a degree of resiliency to the fabric.

The impregnation bath composition is varied dependent upon the end-use of the fabric, and may contain non-flam, antistatic or antisoil emulsions, crosslinking resins, and polyethylene glycol which functions as a humectant plasticiser.

The total solids content of the dispersion as supplied and used may vary between 42 and 50 per cent. It is diluted to give impregnation bath total solids of 20–25 per cent with a dry binder pick-up of between 20 and 25 per cent for interlinings and needlepunch fabrics and 40–50 per cent for underlay fabrics.

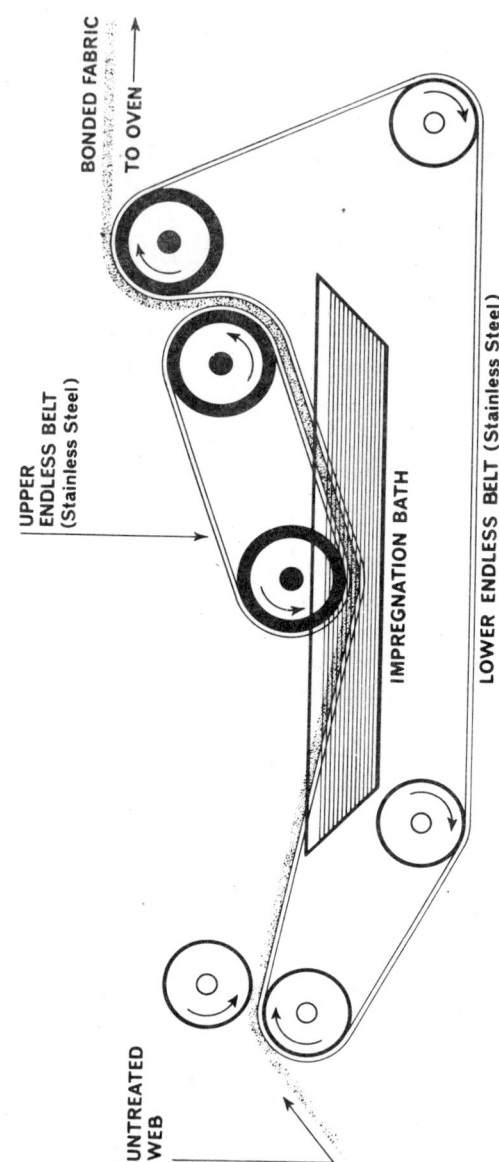

Fig. 6.2. Impregnation bonding.

Spray bonding

The spray bonding of non-woven fabrics is used principally in the manufacture of high loft waddings. This material is used as an inexpensive thermal insulation layer in outer garments, e.g. anoraks.

The fabric webs are prepared by carding the layers and the thickness of approximately 6 mm is maintained. It is the necessity of not compressing the web that determines the method of binder application. Figure 6.3 shows a typical layout of a spray bonding process. High loft waddings are produced at web weights in the region of 0·03–0·05 kg and the binder weight is usually about 20 per cent of the untreated web weight.

In order to impart the resiliency that is required of the wadding, high T_g vinyl acetate–acrylate copolymers are used. These include vinyl acetate–acrylates and styrene acrylates. The binder must exhibit a degree of resistance to dry cleaning solvents and have a high level of water resistance.

The application total solids content of the binder is in the region of 20 per cent and crosslinking resins at a level of 2 per cent dry on dry polymer may be added.

WOVEN FABRICS

In the main, fabrics used for apparel do not receive coating or impregnation treatments. Possibly, the major exception is in rain-wear fabrics where waxes, silicone oils, or the fluoro-chemical resin emulsions, perhaps bonded with a synthetic polymer, may be used.

It thus follows that the principal application of polymer treatments to woven fabrics is in household textiles. Examples of these are upholstery fabrics where waxes, silicone oils, or the fluoro-chemical resin emulsions, used and the compounds derived therefrom can be considered separately.

Upholstery fabrics. These fabrics range from high quality moquettes which require a light backcoating to effect some pile anchorage, and thus assist during the post-dyeing operation, to coatings for cheap cotton base, bulked yarn pile fabrics, which require a backcoating in order to convert them into a handleable product.

For the high quality fabrics, unloaded self-crosslinking acrylic latices of T_g in the range $-24°C$ tend to be used in the compounds. The fabrics may be coated either by roller or knife techniques.

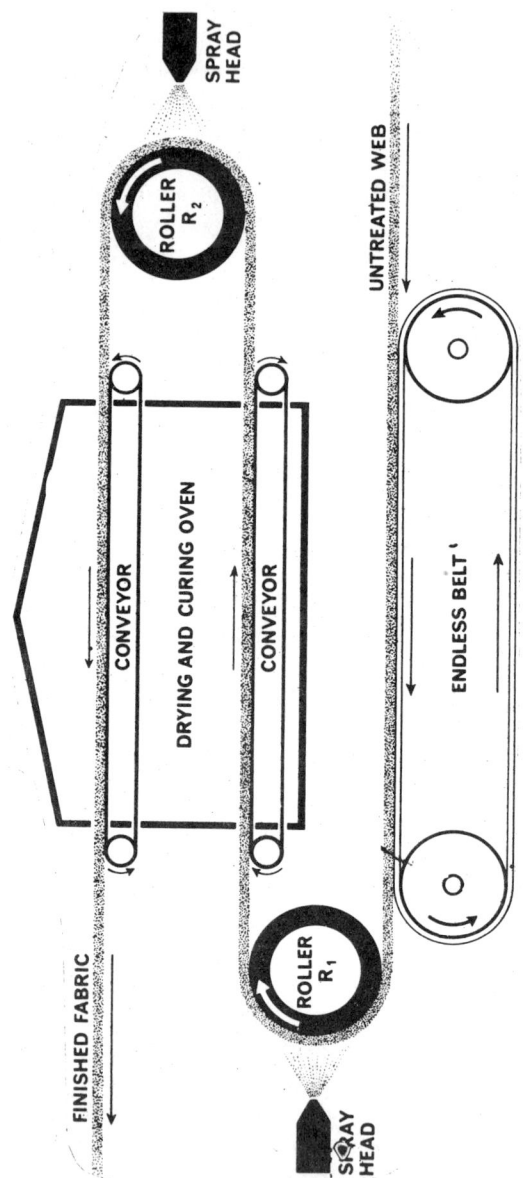

Fig. 6.3. Spray bonding.

A typical recipe for backing moquette fabrics is as follows:

	Parts wet	Parts dry
Acrylic copolymer latex	220	100
Water	108·8	—
Wetting agent	0·8	0·5
Polyacrylate thickener	13·0	3·1
Ammonia solution	2·7	—

TSC 30 per cent, pH 8·5
Viscosity dependent upon method of application

Dry application weight on fabric is 0·05–0·10 kg/m^2. Physical property requirements from the backing include water and 'spot' dry-cleaning resistance, excellent ageing properties and good adhesion to the fabric.

Compounds of interest for the cheaper type of upholstery fabric are usually based on SBR latices and are loaded with a finely ground calcium carbonate. As the fabrics are of low quality, i.e. few ends and picks, backcoating must be performed prior to any subsequent processing. It thus follows that as the fabrics have to be dyed the backing must withstand this process.

The styrene level in the polymers used for this application is approximately 50 per cent with polymer T_g in the range -15 to $-20°C$.

A typical recipe for the compound would be:

	Parts wet	Parts dry
Water	31·0	—
Phosphate dispersant	0·2	0·2
Filler	15·0	15·0
Cure dispersion	2·0	1·0
SBR latex	40·0	20·0
Cellulose thickener	20·0	1·0

TSC 34·5 per cent, pH 8·5
Viscosity 8 Pa.s (Brookfield L at 12 rpm)

Compounds used for this fabric type must be applied by roller coating, as knife coating would cause distortion of the weave. The dry application weight is in the range of 0·12–0·17 kg/m^2. The requirements expected from these backings include:

    adequate spot dry-clean resistance,
    good weave lock (not inherent in the fabric),

a firm but not stiff hand,
a degree of extensibility, and
dye absorption (to give a coloured backing).

The application of back coatings by knife and roller methods has been mentioned. In the roller coating process the fabric is passed over a rotating roller and the compound picked up by the roller is deposited on to the fabric. A 'doctor' knife is applied to the fabric in order to remove excess compound. The application roller is driven at a greater peripheral speed than the linear speed of the fabric in order to achieve an even backing application. After backing, the fabric is passed by means of a stenter through an oven and drying is usually effected at a temperature of between 150 and 190°C.

Knife coating tends to be used where a lightweight application is required on fine (low weight) fabrics. This is achieved by formulating compounds at high viscosity, e.g. 12–15 Pa.s when measured on a Brookfield viscometer. However, knife coating is not used exclusively for this purpose, and any fabric which is not distorted by contact with the knife edge can be backcoated by this method. The knife coating technique in fact allows for greater flexibility in the processing, and by variation in the knife profile, compound total solids content, viscosity and rheology, the degree of penetration of compound into the fabric can be closely controlled.

Bed tickings. Bed tickings are patterned (Jacquard woven) fabric constructed from continuous filament warp yarns, and off-the-loom have poor dimensional stability. A synthetic polymer treatment is required to reduce thread slippage and create a firmer fabric. Plasticised polyvinyl acetate or butyl acrylate vinyl acetate copolymers having a T_g in the range of -5 to $-15°C$ are used.

Window blinds. Fabrics for both roller and louvre window blinds are usually pattern woven, and may be constructed from a wide range of fibre types. The principal requirements of a polymer used in this application are that it gives good water and UV resistance, produces a clear film and imparts stiffness to the fabric without embrittlement.

As is common with many textiles applications, a wide range of polymers can be used to confer the required properties, but price is a limiting factor. Vinyl acetate copolymers containing in excess of 90 per cent vinyl acetate tend to meet the requirements.

REFERENCES

1. CASEY, J. P., *Pulp and Paper*, 1966, **3**, 1551, Interscience.
2. BOOTH, G. L., *Coating Equipment and Process*, Lockwood Publishing Co., London, 1970.
3. *Calendering and Super-calendering*, Lockwood Trade Journal Co., 1964.
4. CASEY, J. P., *Pulp and Paper*, 1952, **2**, 1132, Interscience.
5. STEBBING, T. F. A., *Paper*, 1976, **186**(9).
6. HAGAMASSEY, J., LEE, D. I., SCHMITT, J. A., GIVEN, S. P., HAYES, L. V. Jr., *Tappi*, 1978, **61**(1), 59.
7. SINCLAIR, A. R. (Ed.), *Tappi Monograph*, No. 37, 75.
8. MACARTHUR, M. A., *Paper Technol. Ind.*, January 1978.
9. BUTTREY, D., *Synthetic Polymer Latices in Pigment Coating of Paper and Board*, Revertex Limited Publication, UK, 1975.
10. WELLS CLARK, C., *Blade Coating Technology*, Chapt. 17, Tappi Press, 1978. (Further reading on coating, *Tappi Monographs*, Nos. 17, 25, 28, 30, 36 and 37.)
11. CASEY, J. P., *Pulp and Paper*, Interscience, New York, 1961.
12. LATTIMER, J. J. and GILL, R. A., *Tappi*, 1973, **56**(4), 66.
13. British Patent No. 637 227, *The Bardac Process*.
14. SWANSON, J. W., *Tappi*, 1960, **43**(3), 176A.
15. TAYLOR, P. C., *Paper Technol.*, 1974, **182**(8).
16. SCHOFFMANN, E., *Paper Trade J.*, May 1967, 49.
17. SCHMIDT, S., *World's Paper Trade Review*, May 1972, 656.
18. RADVAN, B. and GATWARD, A. P. J., *Tappi*, 1972, **5**(5), 748.
19. BOOTH, G. L., *Paper Trade J.*, December 1969, 36.
20. WIRA, *The WIRA Carpet Static Load Tester*, details from the Wool Industry Research Association.
21. British Standard BS 4052, October 1972.
22. British Patent No. 856 389, *The 'W' Process*.

Chapter 7

LATEX ADHESIVES

B. A. WOODLEY, J. PRITCHARD and A. A. ARMSTRONG†
Dunlop Semtex Ltd, Birmingham, UK

The use of latex-based or dispersion type adhesives has increased significantly since 1960 and this growth has been due to change in product use, product development and environmental demands. Natural rubber latex at a solids content of 60 per cent plus together with similar polymeric dispersions also have a relatively low viscosity. This means that coatings can easily be applied to a wide range of materials using spray, roller coating or similar techniques, and that high dry coat weights can be achieved from these water-based products.

The viscosities of the same polymers in solution form are high, even at relatively low total solids concentration and, to achieve medium to high dry coat weights, several coats would be required, for example:

	Natural rubber latex	Natural rubber solution
Total solids content	61·5 per cent	17·0 per cent
Viscosity (mPa.s)	150–200	60 000–100 000

In the case of the natural rubber solution, the molecular weight of the polymer has been significantly reduced by milling in order to achieve the moderate solids content.

The high solids/low viscosity relationship means that adhesives based on latex polymers can easily be pumped before and after manufacture, and can be handled with similar ease by the ultimate user. Application machinery, in packaging, textiles, footwear manufacture, etc., can be readily designed to suit the physical properties of such adhesives. Since the continuous phase in these adhesives is water, the absence of volatile

---

† Present address: Du Pont (UK) Ltd, Maryland Avenue, Hemel Hempstead, Herts, UK.

and possibly flammable solvent also enables machinery to be produced without the need for costly flame-proof motors. This is an advantage both to the manufacturers of the adhesive and the company applying the adhesive. Also, no special requirements are demanded in terms of ventilation to remove fumes, flame-proof lighting and segregation from other factory operations.

APPLICATIONS IN THE PACKAGING INDUSTRY

Cold seal adhesives in flexible packaging

The flexible packaging industry consumes thousands of tons of paper and film in the production of packs for confectionery, snack foods, ice cream, biscuits, etc. The majority of these packs are heat sealed, either by utilising a low temperature melt film like polyethylene, or by surface coating the film or paper with a heat seal lacquer. These lacquers may be applied as an all-over coating or pattern.

Some coatings such as polyvinylidene chloride (PVDC) can function as a heat seal layer as well as a very effective moisture barrier coating. Polyvinylidene chloride coatings can be applied as solutions or aqueous dispersions.

To form a bond between coated surfaces, both pressure and heat are required. Heated jaws or rollers on filling machines must run at high temperatures, in order for the heat to penetrate through the paper or film, to soften and fuse the heat seal coating. Thus, whilst the softening point of the heat seal may be 90–110°C, the jaws must be heated to 130–160°C.

Cold seal adhesives containing natural rubber (NR) and other latices have been steadily replacing a number of the applications of heat seal coatings because of certain clear advantages.

   (i) Filling machines, manufactured and supplied without the complicated heating and temperature control systems, cost one-third less in price.
   (ii) Products can be sealed 10–20 per cent faster using cold seal.
   (iii) Reduced power input per machine.
   (iv) Reduction in damaged product, particularly heat sensitive chocolate-based types.
   (v) Less supervision on each machine, i.e. heaters switched on 15–20 min. before starting.

Cold seal adhesives are not new. They came into the market in the late 1960s, mainly from Sweden, and have been used on paper for ice cream wrapping, medical packs and similar applications.

The ability to compound to make an adhesive suitable for use on packaging films has extended their use. The first films used were of the coated type such as MXXT cellulose film and Propafilm C but the increase in the use of co-extruded films has created a wider range of applications for cold seal adhesives.

Most cold seal adhesives are applied by the gravure printing process (Fig. 7.1) since this gives good control for pattern application to the areas to be sealed, and to the coat weight, normally 2·5–3·0 g/m^2.

The performance of a cold seal adhesive is quite demanding. In addition to having good seal strength and film adhesion, the adhesive must have high mechanical stability and non-foaming properties.

Since the majority of packs using cold seal adhesives are for food products, it is necessary for all the ingredients used in the formulation to meet approved standards. Normally the American Food and Drug Administration or the German Bundesgesundheitsamt (BGA) are used, and whilst the adhesive may be pattern applied, direct food contact clearance is necessary. Low dry film odour and freedom from taint are also essential requirements.

Formulations vary depending on the type of surface to be coated, film clarity and sealing pressure used. The general types are:

	Parts (dry mass)
Natural rubber latex	30–70
Resin emulsion	70–30
Stabiliser	1–5
Defoamer	0·5
Antioxidant	1–2

Adhesives for boxes, cartons and cases

Latex dispersion-based adhesives fulfil a demand for improved fast setting formulations to replace long established starch, dextrine and silicate adhesives used in the packaging industry. In many sectors, sophisticated machinery has been developed to use these improved products and give the manufacturer greater output. Using electronic systems, the accurate control of the adhesive application reduces the weight of adhesive required to form the bond. More recently, some of these new adhesives have themselves been replaced by 100 per cent solids hot melt adhesives.

FIG. 7.1. Cerutti gravure printing press. (*Courtesy:* UTP Packaging Company Ltd.)

The three main types of dispersion adhesives used are:

(1) Polyvinyl acetate (PVA) homopolymers compounded and plasticised where required to give flexibility.
(2) Vinyl acetate copolymers including comonomers of acrylate, maleate type and more particularly ethylene. The last copolymer is referred to as ethylene vinyl acetate, although vinyl acetate is the predominant constituent.
(3) Acrylics offer a wide range of molecular weights and, with the various esters available, provide a strong group of adhesives suitable for demanding applications such as bonding to polyolefin films and polyolefin coated board.

Price often controls the selection of adhesive used. Copolymers are more expensive than homopolymers by about 10–12 per cent, whilst acrylics can be 50–100 per cent more expensive than a homopolymer PVA adhesive. The applications of this range of adhesives are wide and complex, but there are some sectors where relatively large volumes are used.

On small cartons made from lightweight board, PVA homopolymer-based adhesives are widely used for side seaming. Application is mainly by roller coating, using the width of the roller to control the width of adhesive strip and a doctor blade, etc., to control the thickness applied.

When erecting cartons prior to filling, PVA homopolymer adhesives are also used alongside hot melt adhesives, as are some natural rubber latex based adhesives. For glue sealing cartons after filling, hot melt adhesives are the most popular, but PVA adhesives have a share of this outlet.

Developments in new spray valves have enabled dispersion-based adhesives to be spray applied under closely controlled conditions. This has given much faster setting speeds and thereby a very significant increase in machine output.

The water resistance of the bonded carton is an important factor. The bottom seal of a detergent carton, for example, requires a strong waterproof bond. Hot melt adhesives or water resistant copolymer acrylics give good bonds to plastic or wax coated board.

The wetting of polyolefin coated cartons can be difficult. Effective compounding is required together with treatment of the olefin surface to bring the surface tension up to above 40 mN/m (dynes/cm).

Similarly, deep freeze packs require effective adhesion and flexibility of the adhesive down below $-40°C$. For less demanding applications, consistent 'tolerant' products capable of trouble free use become more important. Effective 'clean-up' properties also become more important to the carton manufacturers seeking maximum output and less 'down time'.

Cases are made of thicker and often corrugated board. Dispersion adhesives are often used for litho laminating and solid board laminating, although the bulk of adhesives used for solid and corrugated board laminating and lining is still dominated by sodium silicate, starch and dextrine.

Glue lap bonds are made with PVA homopolymer-based adhesives similar to those used on side seaming of cartons, and this is probably the biggest volume in this particular sector of the market since the range and number of cases used is extremely large.

The sealing of cases can be achieved in many ways and often packaging tape is convenient, especially where the number of cases being filled is low, and does not justify an automatic filling and sealing line.

Big volume outlets in the confectionery, food and detergent industries, for instance, require fast automated equipment. Here dextrine, PVA homopolymer adhesives and hot melts are used. The compression time (the time the coated flaps are held in the bonded position) is reduced as more faster setting adhesives are used.

Bottle labelling

The majority of glass bottles are labelled with casein modified adhesives, but as plastic bottles have been increasingly used, new adhesives have been required. Plastic bottles are normally made from polyethylene, polypropylene or polyvinyl chloride, but more recently polyester (PET) bottles have also been introduced.

In most cases dispersion-based adhesives, often copolymer-based with a degree of permanent tack, are used to get adhesion to these 'difficult' plastics surfaces. Flame or discharge treatment assists the wetting of a polyolefin surface.

Envelope manufacture

The worldwide use of envelopes for commercial or private use has led to the development of special envelope manufacturing companies using highly automated machinery. The adhesive or gum used in both the envelope construction and coating the flap for remoistening at a later date, has traditionally been dextrine- or starch-based.

In recent years PVA blends with dextrine have given improved performance, being faster drying and much less prone to cause curl problems. The increased cost of these adhesives has been offset by faster running speeds and product improvement.

A small but important application of adhesive, based on 60 per cent natural rubber latex, is in the production of self-seal envelopes. The particular property of natural rubber which is perhaps unique is that of auto adhesion, or readily sticking to itself, but without having undesirable surface tack and sticking to other materials. This property is fully utilised in the production of self-seal envelopes.

In order to meet the requirements of the envelope manufacturer, the adhesives are compounded to give good adhesion to a variety of paper surfaces, and the viscosity and mechanical stability of each adhesive is adjusted to suit the method of application demanded by the particular envelope making machine being used.

Various machines for envelope manufacture are available, of which Fig. 7.2 is an example. Some use a trough fed roller to apply a controlled weight of adhesive, as a strip, to both the flap and envelope surface. In other types, latex adhesive is fed through a slit, and by mechanically fanning out the envelopes, the surface area presented to the slit is exactly the area of the envelope and the flap that requires covering. The choice of machine depends on the type of envelope being manufactured, and the coating speed is very dependent on the speed of forced drying employed.

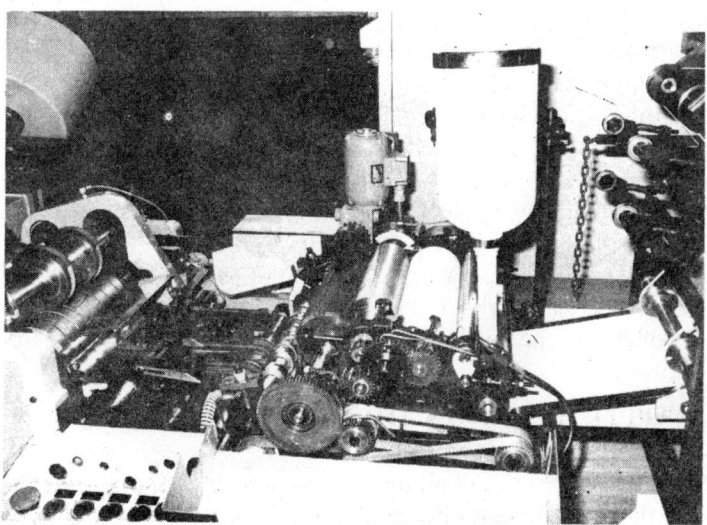

Fig. 7.2. Self seal envelope machine (Winkler & Dünnebier, West Germany).

APPLICATIONS IN THE FOOTWEAR INDUSTRY

Latex-based adhesives have been used in the footwear industry since the early 1900s. Sixty per cent and forty per cent natural rubber latex was used to bond leather to leather in many operations, the waterproof and flexible bond obtained making it an ideal product for use in shoe construction. New materials and automation have changed the types of adhesive to polychloroprene, urethane and hot melt, but natural latex and dispersion-based adhesives are still of value in a number of applications.

The lamination of textile upper fabrics for use in slippers and casuals still uses latex-based adhesives compounded to vulcanise during the final moulding operation. In more traditional footwear, latex adhesives are primarily employed in closing operations (the construction of the top part of the shoe) and in socking (the bonding of a lightweight insole or sock into the completed shoe or sandal).

Whilst brush application is still used, spray and more particularly roller coating are the preferred methods. Roller coating is the convenient way of coating the whole underside of a sock or half sock before sliding it into position in the shoe. The adhesive is only applied to the socking material and there must be sufficient 'open time' in the adhesive to allow the coated sock to be positioned in the shoe before the bond develops. Natural latex and dispersion adhesives are also clean in use, and any excess adhesive getting on the shoe surface is easily removed.

The permeability of one of the surfaces being bonded is important in order to allow the adhesive to set. Some synthetic materials create problems due to lack of permeability.

The dry tack of latex adhesive is very useful in bonding straps, etc., in upper construction, and in manufacturing many bindings and edge trims made by coating, drying and laminating. Again, the waterproof and flexible bond obtained is a very important factor.

Synthetic latex is used in very small quantities. Some polychloroprene latex adhesive is employed to dip polystyrene heels prior to covering, using polychloroprene solvent-based adhesive.

APPLICATIONS IN THE BUILDING INDUSTRY

The use of adhesives in the building industry goes back indefinitely if one considers the use of clay and mud binders for constructing primitive dwellings.

Brick and stone were bonded with a mortar composed essentially of lime, sand and water — the mortar hardening eventually by the conversion of the lime into calcium carbonate. This has stood the test of time in many ancient castles and buildings, which should have encouraged the building industry to accept readily the idea of sticking things together. On the other hand, it would seem natural that it should look for very long life from products used in construction. Quite rightly, the industry is concerned about performance and is therefore conservative in accepting new products.

This position can only be somewhat alleviated by accelerated ageing tests, where long-term predictions have to be taken with a note of caution.

Things are improving nowadays since some polymeric materials have been in use for a quarter of a century and longer, and hence, not only is there an accumulating background of information linking laboratory tests with service life, but also the building industry has evidence of the suitability of polymers from its own experience.

Polyvinyl chloride is now accepted without comment for guttering and associated pipe work and for internal flooring. Polychloroprene and similar polymers are accepted for high quality flat roofing, being much superior to the traditional bitumen.

Wood has been stuck with animal glues for centuries, but in the construction industry any wood glueing would be carried out using the ubiquitous polyvinyl acetate type polymers for faster, cleaner, more convenient work.

Glazing and general sealing between wood and brick, etc., is traditionally done with putties made from whiting and hardening oils, but as the need for sealing became more important because of methods of construction changing, and the introduction of plastics materials, e.g. polyvinyl chloride window frames, then newer materials have been used, among them synthetic latices.

The major uses of adhesives and sealants in the construction industry may be summarised as follows:

(1) Flooring;
(2) Wall cladding;
(3) Ceiling tiles (which may be classed as 'cladding' adhesives);
(4) Sealants;
(5) Wood bonding;
(6) Polyvinyl chloride adhesives for guttering, etc. (semi-structural applications);

(7) Structural adhesives — used in specialised constructions, e.g. laminated wooden beams.

Flooring

As there are many types of floor coverings, so there are and have been many types of adhesives used for this application. Earlier materials such as linoleum were frequently fixed, if required, by resin/alcohol adhesives. Bitumen solutions and solutions of various rubbery polymers have been used with varying success. However, where large areas of flooring are to be laid, using vinyl tiles, the favourite adhesives are water-based, using various types of latex and usually added resin, either as an emulsion or solution.

So far, there are no recognised standards for flooring adhesives. However, there are codes of practice such as CP102[1] and CP203[2] that deal with the preparation of sub floors. This is vital because, as with all adhesive applications, the preparation and type of substrate plays a vital roll in the realisation of a successful and long lasting adhesive joint. Direct to earth floors must be protected with a dampproof course or membrane. Floors must be allowed adequate time to dry and moisture levels are checked where necessary as in CP 203.

Floors should be clean and free from dust, oil, etc. Very dusty floors are best treated with a recommended primer, which is usually based on polyvinyl acetate for cementitious floors. Uneven rough floors would normally be levelled, usually with a proprietary levelling compound, often composed of cementitious materials and a specially compounded latex. Non-rigid floors, wooden floors, etc., are usually covered with hardboard in a recognised method before fixing tiles.

BITUMEN/RUBBER EMULSIONS

Emulsions of bitumen are readily available and can be blended with latices to give adhesives for parquet flooring, vinyl asbestos, linoleum and felt backed floor coverings. These adhesives are comparatively cheap but are black and unsuitable for many applications.

RUBBER/RESIN EMULSIONS

One of the largest uses of adhesives in the building industry is for the fixing of vinyl floor coverings, and the main adhesives used for this purpose are based on blends of rubber latex and resin emulsions. The main type of latex currently employed in this application is based on styrene–butadiene copolymers of one sort or another.

Latex/resin formulations have a lot to offer the user. They are non-flammable, non-toxic, easy to apply, comparatively cheap and usually light in colour. They have another important technical advantage in that they can be compounded to give tacky dry or nearly dry films, which means that most of the water can be allowed to dry off before the tile is finally stuck down.

There are problems associated with sticking vinyl tiles and these can include shrinkage after laying. It does not require a large percentage of shrinkage to cause unsightly gaps to appear between laid tiles. The major problem is one of plasticiser migration from the tile into the adhesive, causing the tile to shrink. A difficulty exists at present in that, as mentioned earlier, there is no standard method of establishing performance criteria. However, the Contract Flooring Association of Great Britain have produced, for guidance, a set of tests for evaluating adhesives for this application.

Rubber latex

Typical adhesive latex properties are as follows:

Styrene–butadiene ratio	50:50
Particle charge	anionic
Total solids content	50 per cent
Specific gravity	1·02
pH	10
Viscosity (25°C)	12 mPa.s

The latices are usually stabilised with soap or surfactant, e.g. 1–4 per cent sodium lauryl sulphate. Antioxidants and preservatives are also added.

Some styrene–butadiene latices are produced with carboxyl groups on the polymer. This is done by copolymerising suitable vinyl acids, for example acrylic, itaconic or fumaric acids. The presence of carboxyl groups imparts excellent mechanical stability, that increases the scope for compounding. Also, it can introduce crosslinking or self cure properties, which helps to combat the effects of plasticiser migration and improve water resistance.

The physical properties of adhesive based on latex depend largely on the film forming properties of the material. Film formation is progressive in that, as the water is lost, the volume fraction of polymer particles in the latex increases. At about 70 per cent volume of particles, interparticle contact occurs, and then particle surface tension and aqueous capillary forces function to minimise surface area. At the same time, polymer molecules begin to diffuse across the particle/particle interfaces, completing the film forming process. As this process is time dependent, adhesive bonds tend to improve with age, unless affected by plasticiser migration. However, it is still essential that no failure occurs in the early stages, i.e. before a bond has reached its final strength.

Resin emulsions

Resins are normally added to styrene–butadiene latex, for flooring adhesive application, to improve adhesion and to impart tack. Obviously, unless water-soluble resins are used, resins can only be added as an emulsion or dispersion. Proprietary resin emulsion/dispersions are available and these can usually be added without difficulty. However, if the compounder wishes to have more scope, he will wish to make his own resin emulsions. Emulsion technology is a subject in itself and, although it is beyond the scope of this book to go into it deeply, a few practical instructions may help the beginner.

The type of resins employed in adhesives can vary a great deal but the common ones include wood resin and its derivatives, mainly esters, hydrocarbon resins, coumarone resin and phenolic resins.

The principles of making a stable emulsion are simple, but in practice a fair amount of experience is required. Basically, the resin has to be converted into very small particles, which have then to be prevented from associating together again.

The production of particles is done either by grinding solid resin in a suitable mill in the presence of water and a suitable surfactant, or by rendering the resin fluid either by melting it or dissolving it in a solvent. Stable emulsions require fairly uniform particle size, usually in the range of 1–10 μm. To improve stability still further, the addition of a protective colloid is usually employed, using such materials as casein, polyacrylates or polyvinyl alcohols.

Where no colloid mill or similar equipment is available, simple emulsions can be prepared with normal laboratory equipment including stirrers. The emulsion will have a large particle size distribution but will be suitable for some formulations. Two examples of emulsion composition (by mass) are as follows:

	A	B
Resin solution		
Resin	40·0	40·0
Solvent	10·0	10·0
Oleic acid	0·7	0·3

Potassium hydroxide solution		
Water	6·0	10·0
Potassium hydroxide	0·17	0·3
Diluents and modifiers		
Water	43·1	24·4
10 per cent ammonium caseinate	—	15·0

The resin solution is made in a suitable container and for the best results should have a viscosity at room temperature of between 2 and 3 Pa.s. The container should not be too big or less shear work is done during the emulsification process. On the other hand, the container has to be big enough to contain the total volume of the finished emulsion with enough free volume to allow vigorous stirring throughout the process.

The potassium hydroxide solution is added slowly to the resin solution with vigorous stirring using a mechanical stirrer. This is the most important stage of the process and, if it is not carried out properly, a poor emulsion may result. The usual fault is to add the potassium hydroxide solution too quickly and/or to generally have too little or inefficient stirring.

Finally, the remaining water or caseinate dispersion is added while stirring.

If a more viscous or more concentrated resin solution (> 40 per cent) has to be used, the emulsion can be prepared at a suitably elevated temperature. Up to 60°C is practicable. If this technique is used, it is vital that the temperature of all the components is maintained at the desired value throughout the process, for otherwise a poor and unstable emulsion is likely. However, the emulsion is allowed to cool to room temperature before mixing it with the latex.

FLOORING FORMULATION

A final formulation suitable for flooring might consist as follows (in parts by mass):

	Dry	Wet
Styrene–butadiene latex (50 per cent)	70	140
Petroleum resin tackifier emulsion (60 per cent)	30–50	50–83
Calcium carbonate filler	50–70	50–70

In Europe, outside the UK more expensive acrylic latices are used. These are less suitable for compounding but can be obtained for wet stick applications when the adhesive dries to a non-tacky very tough film, or they can be obtained as permanently tacky films where water can be allowed to evaporate before bonding takes place.

Wall cladding

The biggest use of adhesive for wall cladding is for fixing ceramic tiles. This is a very old application and originally tiles were stuck with lime-based mortars. Then mixtures of sand and gypsum cement were used up to the 1950s. As more tiles were employed, for example in kitchens and bathrooms, and methods of house construction were altered, more failures were encountered and hence better ways of fixing were required. The newer methods of construction involved the use of non-traditional materials, e.g. breeze block, which gave rise to greater movement of walls.

Some of the earlier adhesives were solvent dispersions of reclaimed rubber with added filler and resin. Those based on black reclaim soon exhibited staining and were replaced by formulations based on so called 'drab' reclaim which is prepared from non-black rubber scrap. Eventually, however, they began to give failures due to embrittlement.

During the 1950s water-based ceramic tile adhesives were introduced. They were based on either rubbery polymers or on polyvinyl acetate latices. Water-based formulations had the advantage of being very easy to use and, unlike their solvent-based counterparts, could be applied to damp surfaces.

A typical early formulation based on styrene–butadiene latex might have consisted of the following ingredients, in parts by mass dry:

Latex (50 per cent)	100
Gypsum	200–300
Tackifying resin	100–150
Surfactant	5–10
Bentonite clay	20–30
Water	To suit

The Bentonite clay imparts a suitable thickness to make the composition trowelable and hence easy to apply. These formulations would also require a fungicide, e.g. sodium pentachlorphenate, to prevent problems in damp conditions. A large filler content is important because of cost, ease of application and the raising of the solids content to reduce shrinkage on drying.

In some instances mixtures of polyvinyl acetate and natural or synthetic latex were blended to form the polymer base. Gradually polyvinyl acetate became the main polymer base for ceramic tile adhesives and now, apart from specialised solvent-based adhesives based on selected cementitious powders, nearly all ceramic tile adhesives are thus based.

Cost has been a constant factor with these adhesives and hence with correct use of fillers and improved polymers, very high filler/polymer ratios are now possible. A typical formulation might include the following components, in parts by mass:

Polyvinyl acetate latex (60 per cent)	50–100
Soluble cellulose	5–10
Filler	600–700
Bentonite clay	10
Water	250

This formulation gives a filler to binder ratio of between 10:1 and 20:1, with a total solids contents of 70–75 per cent. Typical fillers include gypsum, calcium carbonate and silica. Again, the formulation would require a fungicide, e.g. formalin.

Sometimes, coalescing solvents are used to ensure good film formation at low temperatures and to enhance wet bond strength. These coalescents are suitable high boiling solvents such as 2-butoxy-ethanol.

Ceiling tile adhesive
Lightweight expanded polystyrene tiles have become popular and they are usually bonded to ceilings by suitable adhesives. In general, these adhesives are now based on polyvinyl acetate polymers but, in the early days, formulations were sold that were based on styrene–butadiene latices. The polyvinyl acetate adhesives are similar to produce and cleaner in use, being white in colour, and hence accidental marks where they might be seen are less objectionable. When these ceiling tiles were first introduced, a simple way of applying adhesive was to apply blobs at the corners of the tiles and near the centre. The tile was then pressed into position on the ceiling. Adhesives to do this could be thick and buttery. However, due to the behaviour of tiles thus bonded during fires, where the tiles melted and dropped blazing polystyrene onto the floor beneath, adhesive manufacturers issued instructions to the effect that such blob techniques should not be used, but instead an overall coating of adhesive to the whole of the tile surface should be employed. Such new formulations needed to be easier to spread than the early varieties.

Bonding agents

Polyvinyl acetate latices are used extensively in the building industry as general purpose bonding agents and specifically for bonding plaster to difficult surfaces, or for joining new concrete to old.

The use of polyvinyl acetate dispersions for bonding gypsum plaster is considered in British Standard BS 5270.[3] It is primarily intended for those requiring a means of assessing the quality of polyvinyl acetate bonding agents to be used for this purpose. The document includes tests for saponification value, acid value, ash content, bond strength and flexibility.

Factory bonding

Although the uses of adhesives in building so far mentioned have been 'on site' uses, adhesives are employed in some 'off site' applications. There are many advantages in making as much of a building under factory conditions as possible, both from the point of view of weather and general efficiency.

The interest of some years ago in manufacturing 'systems' building for erection on site, has waned somewhat, but there is still interest in 'off site' production of composite laminates, especially for insulation application. Many foam materials in slab form are bonded to suitable skins with polyvinyl acetate and styrene–butadiene latices.

Sealants

With modern building techniques, sealants have become a large and fairly involved topic in their own right. Some sealants are free from a volatile liquid phase and are based on chemically active constituents which cause the sealant to harden after application. Typical examples are silicone sealants and polysulphide sealants. At the other end of the performance scale are what can best be described as 'gunnable putties', i.e. traditional putty-like materials made from a filler and a drying oil, but suitably modified to be able to be extruded from the nozzle of a hand applicator or 'gun'. Somewhere inbetween, both in price and performance, are materials which may have water or a solvent as a volatile medium, which causes the sealant to harden as the liquid evaporates.

The water-based sealants are based on latices of either styrene–butadiene (SBR) or more usually acrylics. A typical SBR sealant formulation is as follows, in parts by mass dry:

Styrene–butadiene latex	100
Calcium carbonate	200
Titanium dioxide	5
Coalescing agent	0–10
Plasticiser	0–10
Wetting agent	1
Ethylene glycol	2

Sealants based on acrylic latices are one-component gun grades and have better ageing properties than SBRs.

POLYVINYL ACETATE (PVA) WOOD ADHESIVES

For centuries the component parts of a large proportion of furniture were assembled by hand using glues derived from animal products. They were used warm, being basically gelatine extracted from bones and hides of animals or fish dissolved in hot water. The advantages of this type of adhesive are fairly low cost and, owing to rapid gelation to a tacky semi-solid on cooling, quick 'grab' properties. However, the disadvantages include limited strength and durability under damp conditions, preparation requirements and a tendency to embrittle with time.

The relatively slow rate of bond stength development further limited their use with the highly automated assembly techniques of the modern production lines. With larger quantities of wood veneers being used together with materials such as plywood, chipboard and blockboard, there was an increasing demand for new adhesives which would rapidly develop high bond strengths between a variety of complex substrates.

Although polyvinyl acetate (PVA) latices have been in existence for about 40 years, they have only been widely used in woodworking applications since about 1955. There are two general types of PVA wood glues:

(1) Conventional PVAs;
(2) Crosslinking PVAs — one and two part systems.

Formulation of conventional PVA wood adhesives
In designing a PVA wood adhesive the following properties will be of major concern:

(i) The adhesive should form a bond stronger than the substrate itself, i.e. wood failure should occur when the bond is tested to destruction;

(ii) The performance and application properties should remain unchanged when used within anticipated working temperatures;

(iii) In many assemblies the bond may need to be resistant to sustained loads, i.e. have creep resistance;

(iv) Even on oak and other woods of high tannin content, the adhesives should not cause staining.

It is vital for the adhesive formulator to select the appropriate PVA latex to enable him to obtain the correct balance of properties. Copolymer latices are, in the main, unsuitable for wood glues because generally they yield soft flexible bond lines which show no creep resistance. Unmodified homopolymer latices do not form a continuous film at temperatures below 15–17°C and, although they could be used as wood adhesives under certain environmental conditions, in Europe at least, they must function satisfactorily at working temperatures as low as 3–4°C, i.e. where wood may be stored outside and brought inside for immediate assembly. It is essential, therefore, that the adhesive is capable of film forming at these temperatures. Normally this would be achieved by the inclusion of a plasticiser, e.g. dibutyl phthalate. However, these should be avoided in formulating wood glues since they produce a more permanent increase in thermoplasticity of the polymer and hence reduce creep resistance. Lowering of the minimum film forming temperature is best achieved by the use of a coalescing solvent or fugitive plasticiser, which is not permanently retained in the adhesive film but only temporarily softens the polymer particles so that film formation can occur.

A polyvinyl acetate polymer, in either latex form or in dried film, will slowly hydrolyse. In latex, acid hydrolysis occurs because the pH is usually between 3 and 5, and a product of this hydrolysis is acetic acid which is an excellent plasticiser for PVA. The reasons for not using plasticisers have been mentioned above, and therefore it is essential that the wood adhesive is buffered at approximately pH 7. This can be readily achieved by the incorporation of a small amount of calcium carbonate, which will not only inhibit hydrolysis but will also protect high tannin content woods against discoloration at low pH.

Having regard for the above points, a basic wood adhesive will consist of a PVA homopolymer latex, a suitable coalescing solvent and a buffering agent, as follows, in parts by mass dry:

Polyvinyl acetate homopolymer dispersion	100
Coalescing solvent	up to 6
Buffering agent, e.g. whiting	2–6
Water (if required)	2–6

The level of coalescing solvent used will depend on its efficiency at reducing the minimum film forming temperature to the desired point and the extent to which it may be retained in the glue line (thereby reducing the creep resistance). Although water miscible coalescents may be directly added to the dispersion, possible precipitation will be avoided if they are first pre-mixed with some water.

Depending on the selection of PVA latex and the type and level of coalescing solvent, the rate of bond strength development of such an adhesive would tend to follow the general curve illustrated in Fig. 7.3. Typical performance figures tested according to BS 3544[4] and BS 4071[5] would be:

Bond strength	1700 N
Resistance to sustained load (446 N)	>7 days

The above formulation would, in general, yield high performance, high cost adhesives with a clear glue line, but it may be modified to meet special requirements by the addition of certain ingredients.

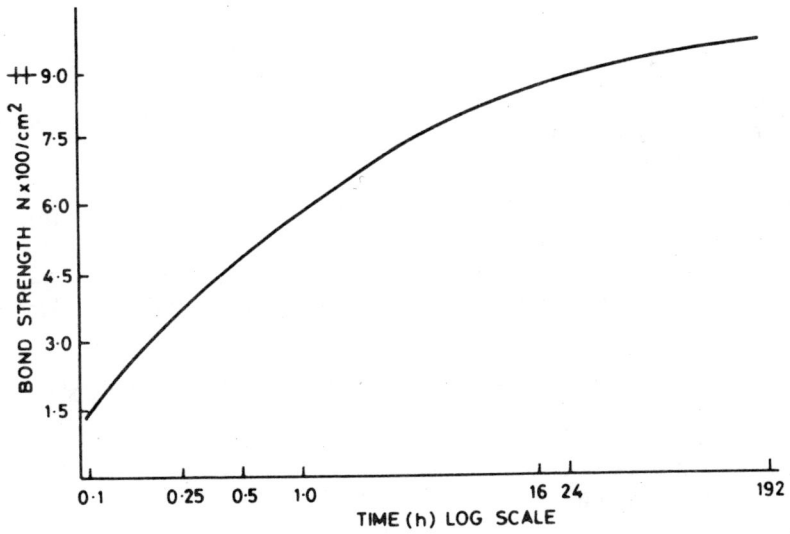

FIG. 7.3. Rate of bond strength development.

TABLE 7.1
The effect of filler addition

Property	Level on dry PVA ≤ 20 per cent	20–100 per cent
Creep resistance	Slightly improved	Lowered
Final bond strength	Unaffected	Reduced
Open time[a]	Unaffected	Decreased
Rate of bond strength development	Slightly reduced	Decreased

[a] Time within which the bond must be assembled to provide satisfactory strength.

Fillers

Table 7.1 illustrates the effect of filler addition. Soft fillers such as whiting and gypsum are preferred because they are less damaging to machine tools than harder fillers such as barytes and limestone.

When costs are of more importance than properties, filler levels of 100 per cent can be used. However, there is the risk of settling and small quantities of bentonite clay or colloidal silica may be included to impart slight thixotropy. Most fillers can be added directly to the PVA latex, but with highly absorbent fillers pre-dispersion in water is recommended.

Thickeners

The final viscosity of the wood adhesive will be governed by the application technique. For hand assembly high viscosity is desirable (10–20 Pa.s) whilst machine work often utilises low viscosity adhesives (2–4 Pa.s). Polyvinyl alcohol is the most suitable thickener and Table 7.2 illustrates its effect on the important properties of wood glues.

Fully hydrolysed polyvinyl alcohols tend to give syneresis on storage of the adhesive and hence the partially hydrolysed types are preferred. Solutions are best prepared by high speed stirring the powder into water at 85–90°C and cooling for about 30 min. The solutions should be

TABLE 7.2
The effect of polyvinyl alcohol on the properties of wood glues

Property	Effect of polyvinyl alcohol
Creep resistance	Increased
Final bond strength	Reduced
Open time	Increased
Rate of bond strength development	Increased

allowed to cool to at least 50°C before adding to the latex. If further economies are sought, then maize starch may be used but only if it is first cooked at about 90°C for 1 h with a fully hydrolysed polyvinyl alcohol, because alone it would lead to loss in viscosity and syneresis during storage.

Prevention of mould growth is an important consideration especially if long-term storage is anticipated. Non-volatile preservatives such as sodium pentachlorphenate, at up to 0·1 per cent on adhesive weight, may be used.

A wide range of wood adhesives can, therefore, be compounded to meet requirements such as:

(i) High quality, high performance;
(ii) Good quality, low cost;
(iii) Fast setting;
(iv) Long open time;
(v) Gap filling (up to 1·3 mm thick).

Depending on these requirements and the service conditions of the assembly, certain performance tests will need to be carried out. In the UK, for instance, two British Standards relating to PVA wood adhesives are BS 4071 and BS 3544, which cover requirements and test methods, respectively, including a freeze-thaw stability test and a staining test, under dry and damp conditions. To conform to these standards double overlap joints, made from European beech (*Fagus sylvatica*), should, in dry conditions, have a failing load of not less than 1334 N and sustain a load of 446 N for seven days. In Germany DIN 68602[6] and DIN 68603[7] specify single overlap shear bonds and do not include a sustained load test.

The latter do not relate to specific 'chemical' types as do British Standards but provide general tests which are intended to indicate the performance to be expected from wood glues of any description. Some harmonisation of European standards, in this area at least, may be achieved with the inclusion of the sustained load test from BS 3544 in DIN 53254,[8] which is intended to replace DIN 68603.

Conventional PVA woodglues are thermoplastic and unsuitable for structural uses and applications where the bond is subject to severe moisture conditions. In these situations thermosetting phenolic resins which in the UK satisfy the 'weather and boil proof (WBP)' requirements of British Standard BS 1204[9] have been shown by extensive experience to be suitable.

Crosslinking PVA wood adhesives

In the 1970s crosslinking PVA latices with improved water resistance were developed. Generally the latices alone are considered sufficiently water resistant to give adhesive joints that meet the requirements of Strain Group B3 of DIN 68602 and DIN 53254.

Further improvements may be obtained through a crosslinking reaction with metal salts such as chromium (III) nitrate or aluminium nitrate, and the subsequent adhesive would attain Strain Group B4 requirements. Typical formulations would be the following, in parts by mass:

	A	B
50 per cent crosslinking PVA latex	100	100
70 per cent chromium (III) nitrate $9H_2O$ solution	5	—
60 per cent aluminium nitrate $6H_2O$ solution	—	5

The aqueous solutions of the metal salts can be readily stirred into the latex and the resultant compound would have a pot-life of about seven days.

Points to note are that the film formation temperature of the modified PVA latex is often sufficiently low to make addition of coalescing solvent unnecessary. Although the latex may be compounded, the acid pH must not be increased and therefore materials of an alkaline nature or containing carbonates must be avoided.

Finally, if a clear glue line is required then compound B, incorporating aluminium nitrate, should be used since chromium (III) nitrate leads to a blue–green coloration in the bond.

Although there are, as yet, no British Standards that apply specifically to these wood glues, BS 3544 and BS 1204 and/or the German Standards mentioned earlier, DIN 68602 and DIN 53254, could be used to evaluate their performance.

Some data relating to the performance of both conventional and crosslinking PVA wood adhesives when exposed to weather and to the conditions prescribed by DIN 68603 have been published by Clad.[10] A study by Beech[11] which compared commercially available catalysed PVAs utilising the test procedures contained in the above specifications and exposure to the weather for 12 months concluded that:

(1) All the glues studied were unsatisfactory for structural purposes or for use in conditions of high moisture where continuous stressing of the joints may occur;

(2) In the absence of further experimental data, the results of the boil

test in BS 1204 should not be used as an indication of the durability of glued joints exposed to weather;
(3) Tests after immersion of samples in cold water for 48 h followed by seven days drying showed greater potential as a predictive test for durability;
(4) Although of generally superior moisture and water resistance to conventional PVAs, tests suggested that crosslinking was incomplete after seven days.

It is therefore important to remember that compliance with such specifications should not be taken as a guarantee that the adhesive will be satisfactory in practice, for use with all types of joints or with all species of timber.

Conventional wood glues are widely used in the preparation of mortice and tenon, plain or bevelled lap joints in both soft and hardwood furniture units, e.g. chairs, cabinets and vee grooving.

In chair frame applications, good gap filling properties are usually required since the pressures applied do not ensure close contact joints. Also, as most chair joints are highly stressed in use, resistance to sustained loads is essential. In cabinet assembly, the continued development of mechanical fixings has resulted in a decline in the use of adhesives. However, the setting speeds and long pot life of PVAs make them the preferred and ideal adhesive for use with automated glue dispensing equipment.

The vee grooving technique has grown in importance (Fig. 7.4). Other angles will produce different unit shapes, while the geometry of the cut determines whether a normal or gap-filling adhesive is required. In some cases, contact adhesives can be used but here the choice depends on the machine and appropriate equipment of the user.

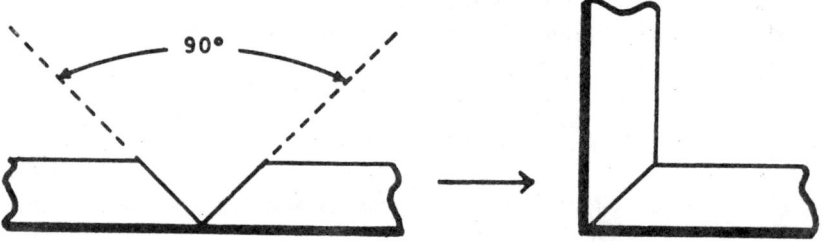

FIG. 7.4. Vee grooving technique.

Other uses are in blockboard manufacture, panel construction and to a lesser extent wood veneering. In this latter area, crosslinking systems could find increasing application since typical flow line press operations bond in excess of 100°C, which would soften ordinary PVAs and lead to blistering when the pressure was released. The use of crosslinkable wood glues for bonding timber for external purposes, e.g. window frames, will increase as practical experience of their suitability grows. If necessary heat or radio frequency may be used to accelerate drying and curing of the adhesive. The adhesive may be applied by brush, serated spreader, roller coating, curtain coating or the popular method of spraying either cold or hot.

In the case of PVA-based adhesives, one surface is coated and assembly takes place whilst the glue is still wet, but contact adhesives rely on the coating of both surfaces and dry lamination under pressure. Bond formation occurs over time as the water in the PVA diffuses into the substrate but no such mechanism operates in the immediate bonding of contact adhesives.

To meet the adhesion demands of laminating other decorative synthetic 'veneers' such as PVC foils, resin impregnated paper, and plastic laminates to less attractive wood, PVA copolymer or acrylic latices are generally used. For bonding PVC foil, the latices alone may be used. However, if plasticiser resistance is of minor importance, then they may be compounded with resins, fillers, plasticisers, thickeners, etc., to give the desired properties.

Fully cured resin impregnated paper can be bonded in a similar manner to that used for wood veneers, whilst the partially cured material uses the resin in the paper as the adhesive. The furniture manufacturers generally bond plastic laminates in platen presses using PVA wood glues, but the trade laminators or shop fitters manufacturing a wider range of panel sizes would just as likely use contact adhesives and a nip roll press. In this latter process latex contact adhesives, which are predominantly based on polychloroprene latex, should gain acceptance.

REFERENCES

1. Code of Practice CP 102, *Protection of Buildings Against Water from the Ground*, British Standards Institution, 1963.
2. Code of Practice CP 203, *Sheet and Tile Flooring, Cork, Lino, Plastic and Rubber*, British Standards Institution, 1961.

3. British Standard BS 5270.
4. British Standard BS 3544, *Methods of Test for Polyvinyl Acetate Adhesives for Wood*, 1962.
5. British Standard BS 4071, *Polyvinyl Acetate (PVA) Emulsion Adhesives for Wood*, 1966.
6. German Standard DIN 68602, *Valuation of Adhesives for Joining of Wood and Derived Timber Products: Strain Groups Adhesive Strength*, 1973.
7. German Standard DIN 68603, *Glued Wood Joints: Testing*, 1974.
8. German Standard DIN 53254, *Testing of Wood Adhesives and Glued Wood Joints*.
9. British Standard BS 1204, *Synthetic Resin Adhesives (Phenolic and Amino Plastic) for Wood*. Part 1 (1964): *Gap Filling Adhesives*. Part 2 (1965): *Close-contact Adhesives*.
10. CLAD, W., Testing adhesives for assembly joints, *Holz als Roh — and Werkstoff*, 1973, **31**, 329–37.
11. BEECH, J. C., The performance of some catalysed polyvinyl acetate (PVA) wood adhesives, *J. Inst. Wood Sci.*, December 1977, **7** (6).

Chapter 8

LATEX PAINTS

K. SELLARS and G. R. BROWN
Harlow Chemical Company Ltd, Harlow, UK

DEFINITION OF A LATEX PAINT

Up to the Second World War and in the immediate post-war period water-based paints were based on natural products such as casein/oil combinations and were known as distempers. In the late 1930s polymer 'emulsions' of vinyl acetate became available and these, together with vinyl acetate/dibutyl maleate copolymers, were used in Germany as the binder for paints. In the UK similar products became available in the late 1940s and were used in the manufacture of emulsion paints. In the USA styrene/butadiene latices were freely available after the war and water-based paints there became known as latex paints. Subsequently, the basis of such paints has been vinyl acetate copolymers and the so-called pure acrylic polymers but the term latex paint has survived. On continental Europe such paints have always been more accurately called dispersion paints. Emulsions are technically a dispersion of one liquid phase in a continuous phase of a second liquid. Since the disperse phase of polymer in a so-called polymer emulsion is a solid and not a liquid, the term polymer dispersion is more correct. Notwithstanding these considerations the term latex will be used rather than polymer emulsion or polymer dispersion.

CONSTITUTION OF LATEX PAINTS

Paints are required to be both decorative and protective. In the decorative context it is essential that the surface is uniformly and completely covered. This is most readily achieved using a pigment of high refractive index such as titanium dioxide, so that the underlying surface is obliterated with a minimum quantity of pigment. It is necessary that the pigment should adhere firmly to the surface and hence a binder is used. Latex paints use latices as the binder. Thus, the simplest, unsophisticated white latex paint could be made by dispersing titanium dioxide in a latex. Unfortunately, most of the properties of modern latex paints would be absent. Opacity would be poor because sufficient pigment could not be added without the latex paint becoming too viscous to apply by brush. Incorporation of a dispersing agent with the titanium dioxide enables the preparation of higher solids, lower viscosity dispersions and is now an essential part of latex paints. Moreover, it permits the replacement of the more expensive prime pigment by other lower cost extenders such as whiting, mica, talc and barytes. Despite this further modification, the latex paint would still be found to be deficient in open time (wet edge) and difficulty would be experienced in making uniform applications to larger areas. This problem can be solved however by the use of a colloid such as a cellulose ether, or more recently alkali-soluble latices, which not only improve open time but also aid flow and levelling. The use of these additives inevitably introduces surface active compounds which promote foam formation, particularly under the brush, and hence defoamers must be added to counteract this. As a further additive a biocide is essential because certain of the ingredients, and in particular cellulose ethers, are liable to degradation by microorganisms. As a final additive it is almost always essential to include a coalescing solvent (temporary plasticiser) to ensure good film formation particularly under adverse conditions such as using the paint at lower temperatures.

The simple latex paint now contains the following ingredients:

(1) Opacifying pigment   (5) Colloid
(2) Latex   (6) Defoamer
(3) Dispersing agent   (7) Biocide
(4) Extenders   (8) Coalescing solvent

Whilst it is firmly contended that simplicity should be the aim in the formulation of latex paints it can be seen that there will inevitably be a degree of complexity.

Before considering in more detail the influence of the various components in the formulation, it is essential to understand the effects of the relative contents of pigment and binder. To achieve this it is important to understand the fundamental concept of pigment volume concentration (PVC) and critical pigment volume concentration (CPVC).

Pigment volume concentration (PVC)
PVC denotes the volume of pigment in a dried paint film and is defined as a percentage.

$$\frac{\text{Per cent}}{\text{PVC}} = \frac{\text{volume of (pigment + extender)}}{\text{volume of (pigment + extender + polymer + other non-volatile ingredients)}}$$

In the early development of latex paints the definition pigment/binder ratio was used to express the amount of pigment plus extender present in the paint. Since no account was taken of the specific gravity of the pigment it was rather meaningless in attempting to compare properties of paints having similar pigment/binder ratios. For example, in some countries paint is sold by weight whilst in others by volume. In the former countries high density extenders such as barytes were favoured on cost grounds and paints with similar pigment/binder ratios could have noticeably different performance.

Critical pigment volume concentration (CPVC)
The concept of CPVC published by Asbeck and Van Loo[1] was that CPVC is the point at which there is just sufficient binder present to satisfy the pigment surface and fill the void spaces between pigment particles.

Theoretically above the CPVC the paint film is considered underbound with resultant changes in paint film properties including increased opacity and permeability. An excellent theoretical model of the calculation of CPVC is given by Bierwagen.[2]

Ramig[3] stressed the interdependence of CPVC and polymer properties and this is referred to later when discussing polymers.

However, for most paint technologists the determination of the oil absorption of the blend of pigments being used is advised. The CPVC can readily be calculated from the practical oil absorption measurement of the pigment blend. It is a good quick guide to the CPVC even though from a purist or theoretical point of view it is not sufficiently accurate.

TYPES OF LATEX PAINT

Having defined PVC and CPVC it is useful to categorise the types of paint in relation to PVC, as given in Table 8.1.

Differing building practices in the various countries dictate the volume and design of exterior paints. In the UK, for example, because brick is used

TABLE 8.1
Types of latex paint

Description	PVC (per cent)	Area of usage	Countries in which predominant	Type of polymer used
Gloss paint	15–20	Interior/exterior trim and woodwork	Western Europe including Spain, Germany and France, USA and Australia	Pure acrylics
Roofing paint	16–20	Mainly sloping roofs	Europe	Pure acrylics, styrene acrylics
Semi gloss	20–25	Exterior wood siding, masonry	USA	Pure acrylics
Silk paints	25–35	Interior walls	Western Europe, UK and Germany	Vinyl acetate/VeoVa 10, vinyl acetate/ethylene/vinyl chloride, some pure acrylics
Wood primers	30–40	Exterior and interior wood, sidings	Europe and North America	Pure acrylics
Primer/UC	40–45	Interior/exterior wood	Europe and North America	Pure acrylics
Matt paints	40–65	Interior/exterior paints (Lower PVC better for exterior exposure)	In most countries	Vinyl acetate/butyl acrylate, vinyl acetate/VeoVa 10, vinyl acetate/ethylene/vinyl chloride, styrene/acrylate
Textured paints	50–70	Exterior masonry, interior walls	Originated in Germany as Putz compositions	Vinyl acetate/VeoVa 10, styrene/acrylate, vinyl acetate/ethylene/vinyl chloride
Matt paints	65–90	Interior only, cheaper quality	In most countries	Vinyl acetate/Veo Va 10, styrene/acrylate, vinyl acetate/ethylene/vinyl chloride

extensively for facing buildings, more than 90 per cent of the latex paint is used for interior decoration.

In order to give practical assistance to paint chemists a series of tested paint formulations is provided in the Appendix at the end of the chapter. Trade names have been used necessarily though reluctantly. Other raw materials of similar type can be used but in many instances modification of the formulation will be necessary in order to achieve adequate stability and optimum performance.

INFLUENCE OF THE LATEX IN LATEX PAINT DESIGN

The types of polymers used in latex paint are very varied and much depends on the state of industrial development and competition within any country. In many developing countries the types available are based on copolymers of vinyl acetate with n-butyl or 2-ethylhexyl acrylate or VeoVa 10†(vinyl ester of branched-chain C_{10} fatty acid). Some styrene/acrylics and pure acrylics are also used. More sophisticated polymers such as those based on vinyl acetate/ethylene/vinyl chloride, which are made in pressure reactors, are mainly confined to the North American continent, Western Europe and Japan. In these areas the market for non-paint products such as adhesives promotes the development of such polymers because such types exhibit specific adhesion to low surface energy substrates (e.g. polythene) which are notoriously difficult to adhere to. In these more advanced economies all the types of polymers mentioned are used.

One of the primary reasons why the various types of polymers are used is the extremely wide tolerance in paint formulation which permits very similar paint properties to be achieved. Moreover, with a given monomer composition it is possible to achieve a wide range of properties by judicious formulation of the aqueous phase. Brendley and Haag [4] have divided water-based products into three categories: (i) aqueous dispersions (latices); (ii) colloidal dispersions; and (iii) water reducibles. Comments which follow are confined only to aqueous dispersions (latices).

Monomer composition
The hardness of the polymer film is of great significance in two respects. First, the performance of the final paint is significantly affected. Second, the unmodified polymer latex does not form an integrated continuous

† Shell trade name.

film below the minimum film-forming temperature (MFT) and this temperature is influenced by the hardness of the film. It is true that the polymer can be made softer by the addition of plasticiser, which at the same time reduces the minimum film-forming temperature, though this may be inappropriate since other properties such as dirt pick-up may be adversely affected. The choice of hardness will, therefore, depend on the ambient conditions under which the paint will be applied and maintained. In tropical countries it is possible to use a harder polymer than would be suitable in a temperate climate. In order to achieve a perspective on the relative hardness of the various polymers, Table 8.2 shows the second order transition temperature (T_g) and the minimum film-forming temperature.

Many of the polymers in Table 8.2 are modified by the inclusion of minor amounts of other monomers such as acrylic or methacrylic acid, and acrylamide to improve freeze–thaw stability, storage stability or wet adhesion to alkyd painted substrates. The level of addition is generally 0·25–4·0 per cent on monomers.

The T_g of the polymers generally lies between $-10°C$ and $28°C$ with MFTs from $+1°C$ to $16°C$. Polymers outside these limits may be used but the bulk lie within this range.

Since T_g is the transition point between brittleness and a soft plastic mass it provides a good indication of the polymer properties. The figures quoted are not absolute and it is known that the rate of heating in T_g determinations can give some variation. The quoted figures were obtained by differential thermal analysis at a rate of $10°C/min$.

The MFT quoted is the temperature at which a crack-free film is produced. Some years ago the MFT was considered to be the same as the white point (the temperature below which the polymer forms a white, powdery deposit). Certain polymer types however show a clear film above the white point. Closer inspection shows the film to contain cracks and hence the MFT gives far better guidance than the white point. The limits of the reliability of both determinations is in the region of $\pm 2°C$. It is of interest to note that the MFT can be influenced by other factors such as water plasticisation and the more hydrophilic polymers containing high levels of vinyl acetate are the most susceptible in this respect. However, water plasticisation can be advantageous in improving film formation. Styrene/acrylics are the most hydrophobic and should inherently display the best water resistance. Nonetheless, careful selection of the aqueous phase ingredients can contribute significantly and recently developed pressure polymers of the VA/E/VC type are equivalent to

TABLE 8.2
Monomer composition

Monomers	Typical combinations	T_g (°C)	MFT (°C)	Typical comonomer range (per cent)
Homopolymers				
Vinyl acetate	Polyvinyl acetate	28	16	—
Styrene	Polystyrene	105	105	—
Vinyl acetate copolymers				
Vinyl acetate/n-butyl acrylate (BA)	80/20	12	7	BA 15–25
Vinyl acetate/2-ethylhexyl acrylate (EHA)	85/15	17	10	EHA 15–20
Vinyl acetate/VeoVa 10[a]	75/25	17	14	VeoVa 15–35
Pure acrylics				
Methyl methacrylate (MMA)/n-butyl acrylate	55/45	10	15	BA 45–60
n-butyl acrylate	40/60	−4	1	
Methyl methylacrylate/2-ethylhexyl acrylate	50/50	−8	5	EHA 40–50
Styrene/Acrylics				
Styrene/2-ethylhexyl acrylate	55/45	9	14	EHA 40–45
Pressure Polymers				
Vinyl acetate/ethylene/vinyl chloride (VA/E/VC)	60/25/15	9	5	—

[a] Vinyl ester of a branched C_{10} fatty acid.

styrene/acrylics in this respect, despite having vinyl acetate levels in the region of 60 per cent.

Plasticisers, usually high boiling liquids such as di-*n*-butyl phthalate, influence both the MFT and T_g. Polyvinyl acetate latex has a T_g of 28°C with an MFT of 16°C. The inclusion of 15 per cent dibutyl phthalate (based on resin solids) reduces the T_g to 0°C and the MFT to 2°C.

Flexibility of the polymer is dependent on the efficiency of the plasticiser or comonomer and for vinyl acetate copolymers the comonomer efficiency is EHA > BA > VeoVa 10 > ethyl acrylate. On a weight basis, the plasticising effect of ethylene is greater than that of other co-monomers capable of reacting with vinyl acetate. The increasing use of VA/E/VC terpolymers for paint is due to lower raw material costs coupled with excellent plasticising efficiency of ethylene and the beneficial effects of vinyl chloride which contributes film toughness with abrasion resistance and alkali resistance. Increasing toughness and hardness permits more ethylene to be incorporated with obvious economic advantages. The initial use of external plasticisers such as dibutyl phthalate gave way to the use of internal plastication where the comonomer was an integral part of the polymer backbone, thus no loss of plasticiser occurred on external exposure and there was, therefore, no film embrittlement. Conversely, high external plasticiser levels in excess of 25 per cent on resin weight have shown remarkable film integrity even after 20 years and even though a high level of plasticiser has been lost.

Flexibility of the latex paint film is important and the degree of flexibility required is governed largely by the substrate. The MFT is less important and can be controlled by the addition of coalescent. Dimensionally unstable substrates such as wood which are exposed to natural weathering require a flexible polymer, for example 50/50 MMA/EHA for primers, whereas a finishing latex gloss paint on the above primer requires a harder polymer such as 55/45 MMA/BA. Highly flexible paints are required for roofs and 40/60 MMA/BA polymers are often used. On masonry finishes, a very wide range of vinyl acetate copolymers, pure acrylics and styrene/acrylics are used and these are normally somewhat harder since masonry is a stable substrate.

Softening of polymers with increasing temperature can lead to impaired dirt pick-up, especially where the polymer is thermoplastic. Latex polymers, however, have high molecular weights in the region of 100 000–1 000 000, and it is this which provides toughness without the necessity to crosslink. Thus, wood primers based on latices have good ageing characteristics, whereas the use of more conventional solvent-based products which crosslink by oxidation often leads to embrittlement. Conversely it creates problems in gloss paints where non-crosslinking during drying leads to poorer mar resistance and inferior blocking properties.

Chemical resistance

Most latex paints in buildings are primarily used for their decorative value on substrates ranging from cementitious rendering, concrete, brick and synthetic fibre boards. A high level of chemical resistance is not generally required.

Alkaline conditions on freshly cast concrete can be severe leading to hydrolysis of some vinyl acetate copolymers. Styrene/acrylics and pure acrylics offer the best resistance to alkaline hydrolysis and vinyl acetate homopolymers and vinyl acetate copolymers of butyl acrylate are the worst. Adequate protection is afforded by copolymers of vinyl acetate with VeoVa 10 provided the VeoVa 10 level is at least 20 per cent. Efflorescence, with consequent attack on the polymer, can occur because of the various salts in brick and lime plasters. In these cases there is likely to be disruption of the film rather than specific attack on the polymer backbone.

Atmospheric pollutants generally cause less of a problem with paint films. The exception is acid produced from sulphur dioxide which leads to attack on paint film components such as calcium carbonate rather than to an attack on the polymer itself. This may lead to permeability of the film thus permitting corrosive liquids to attack vulnerable substrates such as limestone.

In the formulation of latex paints resistance of the polymer to hydrolysis is more important. Most latex paints based on vinyl acetate copolymers have a pH of about 7-8. Working at a high pH in the region of 9-10 is not advisable because hydrolysis of the vinyl acetate can occur, causing the pH to fall to about 7 often with attendant discoloration. Pure acrylics and styrene/acrylics are resistant to hydrolysis and do not have this problem. A much wider pH range is thus possible and this permits the use of alkali-soluble polymers for control of flow and rheology in the latex paint.

Latex water phase

Although aspects of the latex water phase are covered in Chapter 3, it is necessary to consider some of the parameters in relation to the finished latex paints.

Emulsion polymers for latex paints are usually stabilised with anionic and non-ionic surfactants or combinations of the two. A further major stabiliser particularly in vinyl acetate-containing latices is hydroxyethyl cellulose in conjunction with surfactants. The use of colloids such as hydroxyethyl cellulose in vinyl acetate copolymers usually results in significant levels of grafting of the colloid onto the polymer backbone.

Latex paint storage stability and also interactive effects resulting in resin or pigment flocculation are greatly influenced by the latex water phase. Pigment flocculation however can be controlled or minimised by the correct choice of pigment dispersant or by the use of additional surfactants.

The viscosity and rheology of the latex water phase are very significant though they do not necessarily influence the latex paint rheology. In practice many very low viscosity latices produce paints having viscosity similar to those produced from latices having much higher low shear viscosities. Low viscosity latices are naturally preferred for easy handling in paint production.

The latex film should exhibit good water resistance and the latex should be 'robust'. By robust is meant the ability to give stable latex paints in a variety of formulations and where resin and pigment flocculation or destabilisation is minimised. Blanch resistance as an absolute measure of water resistance should be treated with caution because latex particle size can influence this property. A blanched film which remains integrated is obviously much preferred to a film which does not.

Increasing the level of anionic surfactant reduces the particle size of latices whereas the use of colloids generally coarsens the particle size, though neither is a major factor. Particle size does however influence a number of desirable paint properties which are dealt with in the following section.

Latex particle size
Surveys of latices for paints indicate that the majority have peak particle sizes that lie between 0·1 and 1·0 μm. Generally most products have a gaussian distribution of particle size, particularly if the particle size is fairly uniform. Closer investigation of the products indicates that most latices have a peak particle size below 0·6 μm. The particle size is of great

significance as demonstrated below:

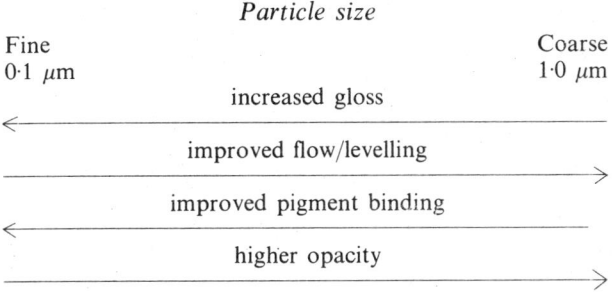

This information is well supported by evidence from the paint industry using standard commercial latices. The latex technologist obviously has a difficult job in achieving the best compromise of properties though means are available whereby properties can be improved against the above trend. For example, toughness can be improved by increasing molecular weight, as also can pigment binding.

In considering flow Kreider[5] demonstrated that the primary factor affecting flow and levelling of latex paints is the stand-off distance, which is the distance between the particles of the latex polymer and the pigment. Contact occurs between particles when the distance between particles is less than the length of an expanded linear substituted cellulose ether molecule. Flow occurs when the rigid network formed by cellulosic linkages is broken under shear. The stand-off distance can be controlled by the particle size of the latex, size of the pigment particles and the volume solids of the paint. The major contributory factor to flow, however, is the latex particle size. Increasing particle size improves flow.

Opacity is an essential factor in latex paints and, *res pares*, is maximised when the difference between the refractive index of the pigment and the polymer particle is greatest. Nonetheless, although the refractive index of the polymer particle plays a less significant role than that of the prime pigment, additional opacity can be obtained by the introduction of air voids which cause further refraction and light scattering. The introduction of air voids occurs more readily above the CPVC and the latter is distinctly affected by particle size. The authors[6] have demonstrated this point by comparing a latex at a particle size of 0·3 μm against a latex having a particle size of 0·9 μm in Formulation 3 (see

Appendix to chapter). By using the larger particle size latex the same opacity can be achieved using 10 per cent less titanium dioxide. This influence of the latex is most significant when the PVC of the paint is above the CPVC (as determined by a practical oil absorption of the pigment blend). The above findings support Ramig's[3] view that CPVC should not be considered solely as a property of pigmentation or latex but a combination of both.

Much work on pigment binding properties has been done by paint technologists worldwide and most rely on the simple test of wet scrub resistance as measured by a Gardener scrub machine or an REL scrub tester. However, great care should be taken in interpretation of such results because many properties are being measured. Such properties include water resistance of the paint, film integrity under the test conditions and adhesion to the chosen test substrate (PVC film, alkyd, hardboard and so on). The basic pigment binding power of the polymer based on such considerations may be difficult to assess. In practical terms, however, one cannot but agree with Nylens and Sunderland[7] who pointed out that there can be no more than a few energetic housewives who attack a newly painted wall with detergents and stiff scrubbing brushes after 24 h, and who scrub the same portion of wall several thousand times.

Undoubtedly fine particle size latices improve pigment binding by virtue of the increased number of particles for a given resin volume provided always that adequate film formation has occurred. Gloss is also improved because of reduced light scattering at the surface.

PIGMENTS AND EXTENDERS

Most latex paints used are white or of a pastel shade. In gloss and semi-gloss paints the pigment is predominantly titanium dioxide whilst in matt paints significant quantities of extenders are included.

A review of the types of titanium dioxide has been produced by Tioxide International.[8] Titanium dioxide can be produced using a sulphate process or, more recently, the chloride process. Of greater significance than the basic production process is the level of coating applied and the post-treatment. The choice of grade depends on the balance of properties required. Lightly coated grades containing 95 per cent of titanium dioxide and having good dispersion properties are used in gloss paints, whereas heavily coated grades containing 82–85 per cent of titanium dioxide are used in matt paints where durability and opacity are usually more critical.

The coating generally consists of alumina/silica and increasing the coating on the titanium dioxide increases the surface area and subsequent water demand and oil absorption. Therefore, for a selected PVC close to the CPVC, high oil absorption pigments give improved opacity because the CPVC is reduced and the PVC is closer to the CPVC. The observations in the section on latex particle size further support this.

High opacity is obtained at low titanium dioxide levels when the PVC is significantly higher than the CPVC. This is because air voids are entrained in the paint film; this is known as dry hiding. Above the CPVC the porosity of the film is increased and this can be monitored by the gilsonite stain test or the enamel hold-out test. The latter involves gloss determination of alkyds applied over the latex paint film.

Other white pigments which contribute to opacity include antimony oxide and zinc oxide. The former is often incorporated to improve flame retardency when used in conjunction with halogenated ingredients. The heat generated by the fire causes the formation of antimony trihalide which prevents air reaching the surface thus retarding flame spread. Zinc oxide is used because of its good fungicidal properties and reduction in the growth of mildew. Great care is needed in formulating latex paints containing zinc oxide because it tends to interact with various ingredients in the latex causing instability. Relatively low levels are generally used but higher levels can be incorporated with latices that are predominantly non-ionic in character. Zinc sulphide and lithopone (zinc sulphide/barium sulphate) are rarely used in latex paints.

Extenders do not directly influence the opacity below the CPVC because their refractive indices are very close to that of the polymer. They must, however, always be considered in conjunction with the titanium dioxide and their inclusion in order to raise the PVC above the CPVC will lead to improved opacity as stated already. Extenders modify packing characteristics which will also affect CPVC. They also influence mechanical properties and durability and cost reduction with extenders can offer both technical and commercial advantages.

Locally mined and graded whitings tend to be the cheapest extenders in most countries and are an important ingredient in most paints. China clay and dolomite are used extensively in Europe and North America because of whiteness.

The water demand of most extenders increases with a reduction in particle size and oil absorption follows the same pattern. A wide variety of particle sizes and shapes are available and these characteristics influence rheology.

Latex paints may be prone to chalking which is a degradation of the polymer mainly by ultra-violet light. In latex paints rutile grades of titanium dioxide are used because of their resistance to chalking though occasionally small quantities of anatase can be included to give controlled chalking and preserve whiteness. The chalking rate of the paint can also be controlled by extenders and it is usually preferable to use talc to modify the chalking rate to attain self-cleaning properties, in preference to a chalking grade of titanium dioxide. Whiting and dolomite tend to reduce chalking but lead to increased cracking. Lamellar extenders such as talc are also useful in reducing or eliminating mud cracking in high PVC paints. Coarse nodular extenders of varying size and shape are employed to produce a range of textured paints.

The natural buffering effect of some extenders, such as whiting, can be exploited in formulation and is usually preferred to the use of chemical buffers.

COLLOIDS

Colloids are water-soluble polymers of varying molecular weight and may be non-ionic or ionic in character. Table 8.3 lists the main types used in the latex paint industry.

Cellulose ether thickeners both of the non-ionic and ionic type are used extensively throughout the world and use of other colloids is currently at a relatively low level. Much effort is being devoted to the development of

TABLE 8.3
Colloids

Non-ionic colloids	Ionic colloids
Methyl cellulose	Sodium carboxymethyl cellulose
Hydroxypropyl methyl cellulose	Sodium salts of carboxyl-containing acrylic ester copolymer
Hydroxyethyl cellulose	
Ethyl hydroxyethyl cellulose	
Polyvinyl alcohol	
Modified starches	
Polyacrylamides	

wholly synthetic colloids based on carboxyl-containing acrylate copolymers or on polyacrylamides.

Colloids influence the following aspects of latex paints:

(1) Control of rheology including flow and application characteristics;
(2) Prevention of pigment sedimentation and maintenance of long-term viscosity stability particularly under hot conditions;
(3) Wet edge properties;
(4) The production of thixotropic (non-drip, gel) paints when used in conjunction with chelating agents.

Latex paint stability and viscosity stability are significantly affected by interactions of the ingredients and relatively little work has been done on colloids in this respect. Kreider[5] has investigated the effects on flow, for example, but in practical terms much is left to the experience of the paint technologist who must assess the properties, often subjectively.

Certain general guidelines are available and, to achieve good flow and easy brushing, the low molecular weight and hence low viscosity grades are selected. Higher levels have to be used to achieve a desired viscosity which is economically less attractive. The wet edge is controlled by water retention and flow. Increasing the level of lower molecular weight material improves water retention and improves flow. The best heat stability is produced by ionic types, particularly sodium carboxymethyl cellulose, though they are often used in combination with non-ionic types because of the susceptibility of the ionic type to bacterial degradation.

Increasing the PVC of latex paints normally requires higher levels of colloid in the finished paint. In many cases up to double the level is used at 70 per cent PVC compared with 30 per cent PVC, assuming that the same grade is employed.

Thixotropic paints are produced by the reaction of titanium chelates such as triethanolamine titanate with hydroxy groups of the cellulose grafted onto the latex polymer. Additional cellulose ether may be added in the paint to give further improvement. The mechanism is not fully understood but it is believed to be due to physical crosslinking caused by hydrogen bonding of the colloid with the chelate in the hydrolysed form.[9] Since this reaction is reversible a thixotropic structure results. The use of polyvinyl alcohol as a colloid or the use of polyvinyl alcohol stabilised latices is not advised because irreversible gels occur. Zirconium derivatives can also be used though closer control of pH is required.

The use of alkali-soluble polymers, generally alkaline solutions of carboxyl-containing acrylic copolymers, is increasing and they are technically important in the formulation of gloss and high sheen (vinyl silk) latex paints. Gloss and flow are significantly improved using such products and performance is greatly dependent on molecular weight, crosslink density and carboxyl content. Care must be taken in the control of pH when using alkali-soluble polymers since paint viscosity and stability can be seriously affected by small changes.

DISPERSANTS

Since nearly all the commercial latices sold are predominantly anionic or non-ionic in character it is not surprising that dispersants follow the same pattern.

A dispersing agent in latex paint is used to deflocculate pigment and extender particles and aid suspension, while at the same time reducing the viscosity of the pigment/extender dispersion. The anion in the water phase is preferentially absorbed onto the pigment particle giving it an electrical charge. Repulsion of similarly charged particles assists in keeping apart dispersed pigment and extender particles provided that the kinetic energy is less than the repulsive charge. In highly viscous latex paints flocculation is reduced because of lower kinetic energies. Obviously the similarly charged anionic or non-ionic latices are unlikely to become destabilised when added to the anionically dispersed pigment dispersion.

Sodium and potassium polyphosphates are widely used as anionic dispersants, as also are very low molecular weight sodium or potassium salts of polyacrylic acid. Dispersion is most effective under alkaline conditions in the pH range 7·5–9·0. The polyphosphate also acts as a sequestering agent. Recently the use of 2-amino-2-methyl-propanol has increased because in addition to being a dispersant it is a very effective pH controller, provided that it is not being used to control polymers that are easily hydrolysed.

Non-ionic surfactants such as nonylphenol–ethylene oxide (20 ETO) adducts are not very efficient *per se* and they are generally used as an adjuvant to the main dispersant to give further improved stability.

Highly stabilised latices are also produced which show good stability even when using only low levels of dispersants in the pigment dispersion. In some cases pigments can even be dispersed in the latex where a high solids paint coating is required. Where necessary wetting agents are additionally added to minimise flocculation of organic pigments or to increase the wetting of difficult substrates.

The optimum level of dispersant for a pigment blend is easily determined by titration of a dispersant solution into a known amount of pigment which is stirred mechanically or by hand. The optimum level is that at which the viscosity of the pigment dispersion is at a minimum. An excessive amount of dispersant can cause flocculation and a viscosity increase in the pigment slurry. The final level of dispersant in the system is affected by the latex used and interactive properties of all the ingredients. It must, therefore, be checked practically in the paint.

Very few problems exist in the dispersion of white and pastel shade latex paints. During the early years of emulsion paint development, most manufacturers produced their own dispersions of inorganic and organic pigments. The situation has now changed and the manufacturers of coloured pigments produce excellent, economic, aqueous dispersions that contain moderate levels of non-ionic dispersants together with propylene or other glycols to prevent drying out and caking.

COALESCENTS

Coalescents (temporary plasticisers) are added to reduce the MFT of the polymer and to assist in film integration. Thus, it is possible to formulate latex paints which can be used at temperatures as low as 5°C even though the latex itself may have an MFT of 12°C or higher. Coalescents are normally high boiling solvents which have boiling points between 185 and 255°C. Typical of coalescents used in latex paints are diethyleneglycol monobutyl ether and its acetate. Consideration of the solubility of the coalescent in the water phase or the polymer phase is a good guide to coalescent efficiency.

Hoy[10] showed that there were two over-riding factors in the performance of coalescents. First, the coalescent should partition principally into the resin phase. Second, the T_g (melting point) of the coalescent should be as low as possible. It is assumed that the T_g of the polymer has been suitably chosen for use in latex paints (see Table 8.2). Additional water-soluble coalescents are included to modify other paint properties such as wet edge and freeze–thaw stability. In gloss paints high levels of propylene glycol are incorporated primarily to improve wet edge. On porous substrates the loss of water-soluble coalescent such as propylene glycol into the substrate can affect film formation.

Although the same principles that apply to latex film formation also apply to latex paint film formation, complications arise because of the pigmentation and, in practice, much higher levels of coalescent are required than would normally be employed to reduce the MFT to a given temperature. In line with this, high PVC paints require a greater percentage of coalescent based on polymer weight than do lower PVC paints. In high PVC paints mud cracking can occur easily and increased levels of coalescent soften the polymer considerably and help to relieve stress cracking.

It is not easy to determine the actual MFT of paints. However, if a coloured paint gives colour development at 5°C equal to that given at 20°C then this signifies that the MFT of the paint is below 5°C. In tropical countries where temperatures above 20°C are normal, coalescents are not required. Good film formation is essential for optimum pigment binding, colour development and paint film performance. The theoretical aspects of latex film formation are well covered by Vanderhoff and Bradford.[11]

Sullivan[12] found that organic solvents leave the film in a surprisingly short time and that the time is dependent on the individual solvent. Solvent evaporation from pigmented films is slowest near the critical pigment volume concentration.

OTHER ADDITIVES

These additives are mentioned only briefly.

Defoamers

Good manufacturing procedures can minimise foaming in latex paints but minimal levels (0·05–0·3 per cent) of defoamers are incorporated to eliminate foaming during application by brush or roller.

Biocides

It is always advisable to include an in-can preservative. Preservation of the paint film is highly desirable under tropical conditions, though even high levels of organic biocides and fungicides tend to be shortlived because of leaching. Inclusion of inorganic fungicides such as zinc oxide or barium metaborate is more permanent.

Good housekeeping in production areas can eliminate yeast and bacterial contamination. Regular cleaning down and soaking tanks with a solution of sodium hypochlorite is a valuable exercise. Regular checks for bacteria should be made on susceptible raw materials such as whiting and aqueous dispersions of organic pigments.

Inorganic thickeners
Naturally occurring Bentonite clays (Montmorillonite) and Hectorite are sometimes used to increase viscosity and thixotropy in latex paints.

DEFICIENCIES OF LATEX PAINTS

Latex paint rheology
It is beyond the scope of this chapter to discuss in detail the rheology of latex paints. Suffice to say that almost all latex paints are pseudoplastic or thixotropic, in contrast with solvent-based paints which usually have Newtonian characteristics. The consequence is that whereas the brushmarks produced with solvent-based paints level out and disappear, the brushmarks from latex paints remain visible to a greater or lesser extent. Whilst many factors affect the flow of latex paints, it can be readily demonstrated that the use of low particle size (0·2 μm) latices, which are usually surfactant stabilised, leads to significantly poorer flow than larger particle size (0·8 μm) latices, which are normally stabilised by colloids. Much further work is needed in this area since even the best flowing latex paints fall short of an average solvent-based product.

The deficiency is more significant in gloss paints because the brushmarks detract markedly from the appearance. In countries like the UK, where there is a high usage of solvent-based alkyd paints for wood trim, it is unlikely that a significant market would be created for gloss latex paints unless flow were markedly improved.

Adhesion
Adhesion of latex paints either to themselves or to stable, slightly porous, substrates is not a problem. On friable or highly porous masonry substrates a sealer is required. However, difficulties occur when latex paints are used over old alkyds, chalky surfaces and mild steel. In the case of old alkyd substrates the major problems occur under wet conditions and much work has been done to overcome this. The problem can be eliminated by inclusion of a wet adhesion promoter into the polymer backbone, as described later in the chapter. Their inclusion in the latex adds significantly to the raw material cost and the authors

retain some doubts as to whether the incidence of such failures justifies the additional cost solely for that purpose.

Poor adhesion to chalky surfaces is caused by the lack of penetration of the substrate by the latex. Provided any loose surface is removed the problem can be overcome by the use of a solvent-based sealer or, more recently, a very fine (0·02–0·04 μm) particle size latex. The inclusion of alkyd emulsions into the latex paint is also said to give improvements though ease of brushing is usually sacrificed.

Adhesion to mild steel substrates often relates to a failure to make intimate contact with the surface, especially when contaminants such as oil are present. Though these contaminants may be present only as a monomolecular layer, adhesion can be seriously reduced.

Thermoplasticity
Polymer particles in a latex film do not crosslink and hence the film remains thermoplastic. Alkyd paints on the other hand crosslink by auto-oxidation and hence the film is relatively non-thermoplastic.

Thermoplasticity is particularly significant in low PVC paints such as gloss or semi-gloss where the resin content is high and the surface is resin rich. These paints generally show poor blocking characteristics and mar resistance is poor. This is covered further in the section below on Latex Paint Developments.

LATEX PAINT DEVELOPMENTS

Gloss paints
Latex gloss paints have always been the goal of the paint technologist. The almost insuperable problems of obtaining good wet edge and flow, exacerbated by variable humidity conditions coupled with film thermoplasticity, have prevented the development of latex gloss paints giving comparable performance to solvent-based alkyd gloss paints.

When compared with an alkyd gloss paint, even the best of available latex gloss paint systems show reduced glass and inferior flow, wet edge and thermoplasticity. Significant breakthroughs will be required to achieve radical improvements, although the use of alkali-soluble colloids has gone some way to improving flow. Thermoplasticity can be reduced by the use of latices which crosslink under ambient conditions although current means of achieving such results are either totally uneconomic or lead to other undesirable properties such as yellowing.

Matt paints

Matt paints cover the PVC range 40–80 per cent and significant improvements in performance are not readily envisaged. In recent times high inflation coupled with trade recession has concentrated effort on cost reduction with retention of performance. It has been noted that at 50 per cent PVC approximately 85 per cent of the raw material cost of the paint is that of the latex and titanium dioxide. Even at 70 per cent PVC this cost is about 75 per cent. The quantity of latex often cannot be reduced without affecting performance. The quantity of titanium dioxide, however, can be reduced by using a latex conferring increased opacity provided that the formulation lies in the appropriate PVC range. Alternatively, total solids of the paint could be reduced but this leads to undesired side effects. Quite recently so-called plastic pigments have replaced titanium dioxide and Glidden[13] have produced polystyrene 'pigments' and Ciba Geigy[14] have produced 'pigments' based on urea–formaldehyde resins.

The use of vinyl acetate/ethylene/vinyl chloride latices continues to increase in Europe and types are now available to cover latex paints throughout the PVC range, as well as more specialised latices for putzes and high build finishes. Although basic raw material costs of these latices are lower than other types giving comparable performance, the large investment both in pressure plants and the essential research and development governs prices in the market place.

Flame-retardant paints

Flame-retardant paints have been known for many years. Intumescent types are based on ammonium dihydrogen phosphate or ammonium polyphosphate together with other specialised ingredients such as tri-pentaerythritol and chlorinated paraffins. Non-intumescent types are based on antimony oxide in combination with halogenated compounds. It is likely that such paints will assume an increasingly important role and there has been recent work on the use of styrenephosphonic acid. The VA/E/VC latices should also be of interest because of their chlorine content.

Improved adhesion

Adhesion may be a problem as noted previously and in the case of adhesion to alkyd substrates under wet conditions special monomers have been introduced into the polymer backbone. Such monomers include glycidyl methacrylate,[15] allyl acetoacetate[16] and unsaturated ureido compounds.[17] Wet adhesion can also be obtained by the reaction of ethyleneimine with carboxyl groups on the polymer backbone.[18] It is possible that different monomers could be designed to improve adhesion to other substrates. In more general usage, adhesion to chalky substrates is a problem and products have appeared on the market which claim to give significant improvements.

Other additives

It was mentioned earlier that the use of alkali-soluble products is increasing. Careful design of the polymer can lead to such diverse effects as improved flow, improved dispersion and stability, and highly efficient thickening. It seems likely that such products or others of similar chemical constitution may eventually replace a far larger part of the cellulose ether market. Additives to reduce thermoplasticity may also be developed though, at the present stage of technical development, this is more likely to be achieved by incorporation directly into the latex polymer backbone.

Pigment/resin technology

Developments are likely in the future by combining the technologies associated with latex production and pigment production. The latex manufacturer has produced new developments at the expense of the pigment manufacturer. The pigment manufacturer has introduced new pigments such as Spindrift, a vesiculated titanium dioxide from Tioxide International.[19]

A problem of paint formulation is not only the individual separation of all the pigment particles by excellent dispersion but also a means of retaining this separation in the dried paint film. In practice this never occurs. If, therefore, the latex polymer could be introduced onto the pigment particle, so that the pigment particle became the core and the polymer the shell, then there would be a significantly greater likelihood of achieving better opacity with good film properties. Despite various patents there have been no products of this type on commercial offer.

REFERENCES

1. ASBECK, W. K. and VAN LOO, M., *Ind. Eng. Chem.*, 1949, **41**, 1470.
2. BIERWAGEN, G. P., *J. Paint Technol.*, 1972, **44**(574), 46.

3. RAMIG, A. Jr., *J. Paint Technol.*, 1975, **47**(602), 61 and 64.
4. BRENDLEY, W. H. and HAAG, T. H., *Non-polluting Coatings and Processes*, J. L. Garden and J. W. Prance (eds.), Plenum Bros, NY.
5. KREIDER, R. W., *Offic. Dig.*, 1964, **36** (478), 1244–60.
6. BROWN, G. R. and SELLARS, K., *The Use of High Opacity Emulsions in Paints*, Harlow Chemical Co. Publication, April 1980.
7. NYLÉN, P. and SUNDERLAND, E., *Modern Surface Coatings*, Interscience Publishers, New York, 1965, 662.
8. Publication BTP164 Part 1, Tioxide International, UK.
9. Publication TIL10, Titanium Intermediates Ltd, UK.
10. HOY, K. L., *J. Paint Technol.*, 1973, **45**(579), 56.
11. VANDERHOFF, J. W. and BRADFORD, E. B., *Tappi*, 1963, **46**(4).
12. SULLIVAN, D. A., *J. Paint Technol.*, 1975, **47**(610), 60–7.
13. US Patent 4,069,186.
14. RENNER, A., *Coating*, 1979, **10**, 254.
15. British Patent 1,483, 058.
16. German Patent 2,535,372.
17. US Patent 3,300,429 and British Patent 1,104,344.
18. British Patent 1,088,105.
19. British Patent 1,288,583.

Appendix continues on page 168

APPENDIX: FORMULATIONS

Formulation 1

LATEX GLOSS PAINT

Item	Mass
1. Revacryl 451 (40 per cent TSC)	3·0
2. Water	2·8
3. Ammonia (0·880)	0·3
4. Propylene glycol	7·5
5. Texanol	1·1
6. Hercules defoamer 1512 M	0·2
7. Titanox RA 61	19·4
8. Revacryl 301 (51 per cent TSC)	55·0
9. Water	8·3
10. Revacryl 450 (40 per cent TSC)	2·1
11. Ammonia (0·880)	0·3
Total	100·0

Solids content	50 per cent
Volume solids	40 per cent
PVC	15 per cent
Specific gravity	1·23
Final pH	9

Alkali-soluble polymer
as percentage of total
binder:

Revacryl 301	94 per cent
Revacryl 451 dispersant	3·5 per cent
Revacryl 450 thickener	2·5 per cent
Gloss at 20°C	55–60 per cent
Gloss at 60°C	85–90 per cent

Manufacturing procedure
(1) Make alkaline solution of Revacryl 451 by mixing first three ingredients;
(2) Add items 4–7 and disperse;
(3) Let down with Revacryl 301 followed by water;
(4) In production it may prove better to add the ammonia before the Revacryl 450 to get a smooth transition in viscosity.

Formulation 2

VINYL SILK LATEX PAINT

	Mass
Water	22·6
5 per cent Tetron solution	0·9
Orotan 850	0·5
Bevaloid 691	0·1
Ammonia (0·910)	0·1
3 per cent Natrosol 250 MR solution	13·8
Tioxide R-HD2	27·5
Emultex VV536	33·0
Texanol	1·4
Proxel CRL	0·1
Total	100·0

Solids content	46 per cent
Volume solids	30 per cent
PVC	30 per cent
Specific gravity	1·31

Viscosity:
  ICI Gel Tester   55 g/cm
  Brookfield L (4/30)   7·2 mPa.s
  Stormer   92 KU's

Gloss (on glass):
  (Sheen) 20°C   47 per cent
  (Sheen) 60°C   75 per cent

Formulations 3 and 4

MATT LATEX PAINTS	3 Interior/Exterior (Mass)	4 Cheap interior (Mass)
Tioxide RXL	20·0	11·0
Snowcal 6ML	11·1	29·0
China clay M100	5·0	—
Speswhite	—	5·0
10 per cent Calgon PT solution	—	1·0
5 per cent Tetron solution	2·0	—
Dispex N40	—	0·5
Orotan 850	0·5	—
Bevaloid 691 (defoamer)	0·2	0·1
3 per cent Methocel J12MS solution	—	16·0
3 per cent Bermacoll E270G solution	18·0	—
Ammonia (0·910)	0·1	0·1
Butyl diglycol acetate	1·0	—
Texanol	—	0·5
Dibutyl phthalate	—	0·25
Proxel CRL (preservative)	0·1	0·1
Water	17·1	26·45
Emultex AC430 (TSC 55 per cent)	24·9	—
Emultex VV536 (TSC 55 per cent)	—	10·0
Total	100·0	100·0
Solids content	51 per cent	51 per cent
Volume solids	23·4 per cent	29·4 per cent
PVC	*50 per cent	75 per cent
Specific gravity	1·4	1·43

*CPVC 47 per cent

Notes on formulations

LATICES
Emultex VV536:	VA/VeoVa 10 copolymer —	Harlow Chemical Co. Ltd
Emultex AC430:	VA/BA copolymer —	Harlow Chemical Co. Ltd
Revacryl 450:	Acrylic alkali-soluble thickener —	Harlow Chemical Co. Ltd
Revacryl 451:	Acrylic alkali-soluble dispersant/gloss promoter —	Harlow Chemical Co. Ltd
Revacryl 301:	Pure acrylic —	Harlow Chemical Co. Ltd

TITANIUM DIOXIDES
Tioxide RXL	— Tioxide International
Tioxide RHD2	— Tioxide International
Titanox RA61	— Kronos Titanium Pigments Ltd

EXTENDERS
Snowcal 6ML:	Whiting —	Blue Circle Enterprise
Speswhite:	China Clay —	ECC International
China clay M100:	Calcined China Clay —	ECC International

DISPERSANTS
Tetron:	Polyphosphate —	Albright & Wilson Ltd
Calgon PT:	Polyphosphate —	Albright & Wilson Ltd
Dispex N40:	Sodium polyacrylate —	Allied Colloids
Orotan 850:		— Rohm & Haas Ltd

COLLOIDS
Natrosol 250 MR:	Hydroxyethyl cellulose —	Hercules Power Co. Ltd
Bermacoll 270G:	Ethylhydroxyethyl cellulose —	Berol Kemi (UK) Ltd
Methocell J12MS:	Hydroxypropyl-methyl cellulose —	Dow Chemical Co. Ltd

ADDITIVES
Texanol:	Coalescent —	Eastman Chemical International AG
Butyl diglycol acetate:		Hoechst AG
Bevaloid 691:	Coalescent —	Bevaloid Ltd
Proxel CRL:	Defoamer —	ICI Ltd
	Preservative -	

Chapter 9

LATEX DIPPING

D. M. Bratby
LRC Products Ltd, London, UK

Latex dipping processes appeared as natural rubber latices became commercially available in an adequately stable form in the period around 1930. They followed on as a development of rubber solution dipping because they were less hazardous, more economic and technically more versatile with their high rubber content at a low viscosity. Equally, compounds prepared for latex dipping appeared to meet the requirements of a simple manufacturing process for the production of seamless rubber articles. Unfortunately, successful exploitation using this technique has not proved to be so simple or as easy.

As it was anticipated, the art and technology of latex dipping has progressed to its current state with the technology still following the art in many areas. There are signs that basic research into these processes is increasing both in the colleges, research institutes and commercial organisations particularly associated with latices and dipping. It is still an area that will prove fruitful in terms of discovery for the foreseeable future. Not least because it involves dealing with the bulk properties of complicated colloidal systems which in the production sense perform satisfactorily and then 'go wrong' if inadequate controls are applied to them.

This is an indictment of our fundamental lack of understanding of the problems that confront a latex technologist in a manufacturing or research situation. The need for clarifying the basic technology and knowledge to enable latex dipping to progress out of the art and move into the science is essential.

Despite this background, many successful processes in the latex dipping industry are operating today, manufacturing such articles as:

(1) Prophylactics;

(2) Gloves — surgeons, houseware (unsupported and fabric lined), electricians and heavy industrial types;
(3) Balloons — novelty, advertising, meteorological;
(4) Catheters of various types;
(5) Baby feeder teats and soothers;
(6) Seamless football bladders;
(7) Wellington boots and over-shoes;
(8) Castration rings for animals.

Essentially the latex dipping process consists of using an inert former or mould, which is in the shape of the ultimately desired product, and coating it with one or more dips of the latex compound. The coating is set by a coagulant and/or heat, and dried into a continuous film which can then be stripped from the mould. It is the purpose of this chapter to elaborate on this sequence of events, and the art and technology related to them in commercial processes.

THE FUNDAMENTALS OF DIPPING AND MOULD DESIGN

The solid/liquid interface

The uniform spreading of a liquid over a solid former is an essential requirement for successful latex dipping. This is achieved when the angle of contact between the liquid and the solid surface is zero. When it approaches 180° a total non-wetting situation has developed, which reduces as the angle of contact falls.

If unsatisfactory wetting occurs in a dipping process, then an irregular latex deposit is formed with at best unevenly spread latex, or at worst a complete break-up and separation of the film. There are two factors relating to these effects:

(i) Surface energy of the solid dipping former;
(ii) Surface free energy of the latex.

For complete wetting to occur (i) must be higher than (ii). From this theory it can be predicted which types of materials would make the best dipping formers, e.g. porcelain, glass, metal, as opposed to low surface energy materials like plastics. Despite the theoretical reasoning, some plastic formers are manufactured, e.g. polyethylene, polypropylene and PVC, because the surface free energy of latex compounds can be reduced to low values by the addition of very efficient surfactants.

Vertical dipping

From the later text it is apparent that most dipping processes utilise vertical dipping techniques which would be unacceptable if the contact angle, θ, for wetting was greater than zero. The net result of $\theta >$ zero would be the trapping of air on the former surface at the point of entry in the latex and dragging it down into the latex compound.

If $\theta =$ zero and the downward dipping velocity exceeded the rate of wetting, air inclusion could still occur.

Former design

Where a former touches the latex surface it should be preferably in a vertical attitude. This is obviously not practical in a contoured former, e.g. a glove shape without some undercut or rounded shape being introduced onto the former. Taking the example of this shape, certain basic rules apply:

(i) All surfaces entering the latex that are undercuts must be smoothly pointed, e.g. finger tips, finger crotches. This will minimise air bubble trapping.

(ii) All protruding undercuts must be angled to blend into the hand. This particularly applies where the ball of the thumb section is joined to the palm.

(iii) Any roughening pattern on the former can only be embossed as deeply as the latex or coagulant deposit is able to successfully cover it without break-up.

(iv) Roughening patterns must be orientated in the direction of dipping. This will feed the air out of a correctly designed profile instead of being caught in an undercut.

The most common material for latex dipping former manufacture is porcelain. Its advantages are:

    (i) Can be formed into complicated shapes with mould lines wiped clean prior to firing the ceramic;
   (ii) Patterns can be embossed into the basic mould with high definition;
  (iii) The overall shape and pattern can be sprayed with fine slip coating or grit to soften the pattern and improve the coagulant or latex pick-up;
  (iv) The material has a high heat capacity;
   (v) It is resistant to chemical attack;
  (vi) It can be made extremely tough to mechanical shock, and if

damaged usually breaks so that it can be seen and replaced;
(vii) It has adequate heat resistance to survive a thermal shock of up to 60–80°C without breakage.

LATEX COMPOUNDING

Natural rubber latex

Of all the polymer latices that are used for dipping processes, natural rubber is the most important, particularly in non-specialised applications. There are several important features which must be recognised in this natural product, which are:

(i) A high total solids latex of 60 to 70 per cent that is ammonia or fixed-alkali stabilised.

(ii) The latex is readily available in high-ammonia and low-ammonia forms which permits deliberate stability control, necessary in latex dipping. Colloidal stability can be enhanced by the addition of soaps and surfactants, or reduced by the addition of zinc oxide. Stability can also be reduced by lowering the ammonia level of the latex by neutralisation with formaldehyde, or by heat expelling, or by flushing ammonia from the latex by bubbling air through it.

(iii) It has an excellent wet gel strength.

(iv) It can be prevulcanised prior to use.

(v) It can be straight, coagulant, or heat-sensitised dipped. Even a porous former will produce a coherent film, particularly if it contains calcium ions, which destabilise the latex.

(vi) A rapid drying and curing characteristic is typical, unlike most synthetic latices.

Compounding aims

There is a conflicting requirement for the colloidal stability of latex during its preparation prior to being used in a manufacturing process. During the compounding of the latex it is preferable to maintain a state of high colloidal stability which must later be reduced for the dipping process. It is the very essence of dipping (particularly coagulant or heat-sensitised dipping) that the compound will be sufficiently unstable in order to deposit an adequately thick rubber film, although much depends on whether prevulcanised or unprevulcanised compounds are used.

Unprevulcanised latex compounds

It should be understood that unprevulcanised latex compounds are only unprevulcanised in the sense that they have not been deliberately heated to achieve a state of cure in the polymer. It would be incorrect not to recognise that an inconsistent dipping compound will result from an unmatured, fully compounded natural rubber latex.

This means that a minimum period of 1–3 days should elapse after compounding before the latex is dipped. This enables the compound to free itself of air entrained during the preparation and allows the stabilisers to distribute themselves uniformly throughout the aqueous and dispersed medium.

During this maturation period important changes take place. Absorption of vulcanisation ingredients onto, or into, the rubber particles surface commences and becomes a continuing process with time and temperature.

Natural rubber latex is unique in its reaction with zinc oxide or zinc carbonate, and the technology of latex dipping associated with the chemical changes involved forms a very important part in the control and utilisation of natural rubber latex for dipping processes. A comprehensive discussion of this effect can be found in the literature.[1] Typical formulations are given in Table 9.1.

Prevulcanised latex compounds[2]

Fully prevulcanised natural rubber compounds are stable latices which develop significant strength properties by low temperature drying. These properties can be improved by extra vulcanising ingredients and additional heat. However, the compound will lose its inherent latex cure state stability, and the user has to balance the manufacturing process application against the extent of the material modification.

Synthetic latices

Although the processes that will be described are all applicable to natural rubber latex dipping, some of them are equally applicable to synthetic latex dipping, particularly with nitrile and polychloroprene latices. These are speciality materials designed to dip satisfactorily. Certain aspects are important and characteristic of each of these polymer latices.

TABLE 9.1
Unprevulcanised latex compounds

Typical formulations	Houseware glove no. 1 dip		Flock adhesive houseware glove no. 2 dip		Surgeons glove		Prophylactics		Balloons	
	Dry	Wet	Dry	Wet	Dry	Wet	Dry	Wet	Dry	Wet
60 per cent natural rubber latex	100	167	Not recommended with an unprevulcanised latex		100	167	100	167	100	167
10 per cent potassium hydroxide solution	0·50	5·00			0·5	5·00	0·5	5·00	0·5	5·00
10 per cent potassium oleate solution	0·25	2·50			0·25	2·50	0·25	2·50	0·25	2·50
50 per cent sulphur dispersion	1·50	3·00			0·75	1·50	1·00	2·00	0·5	1·0
50 per cent zinc diethyl dithio-carbamate dispersion	1·00	2·00			1·00	2·00	—	—	—	—
50 per cent zinc dibutyl dithio-carbamate dispersion	—	—			—	—	0·75	1·50	0·75	1·50
50 per cent 2,2' methylene bis(4-methyl 6-tert-butylphenol) dispersion	1·00	2·00			1·00	2·00	1·00	2·00	1·00	2·00
50 per cent zinc oxide dispersion	1·00	2·00			1·00	2·00	1·00	2·00	1·00	2·00

N.B. For surgeons glove compound it is important to omit potassium hydroxide if inadequate leaching is applied to the finished product as this causes skin sensitisation for the end-user. In this case a non-ionic stabiliser and ammonium caseinate will provide sufficient additional colloidal protection.

Nitrile latices

Typically nitrile latices vary in acrylonitrile content from 18 to 42 per cent of the total polymer. They are almost always modified by a degree of carboxylation introduced into the polymer chains. These two factors have a significant role in determining dipping behaviour and properties of the finished product.

As the acrylonitrile content is increased, the less resilient does the gelled latex become, and its degree of tack is diminished. With increasing acrylonitrile content the finished product will be more plastic, having greater solvent resistance, and modulus increases as the polarity of the polymer increases. The advantage of a non-sticky rubber in the finished product has the compensating disadvantage that the polymer cannot be beaded satisfactorily. The formation of a bead would be a technical advantage for the glove in service, as the low tear strength enhances tearing at the cuff edge.

The degree of carboxylation has significant effects on the latex dipping performance and on the finished product characteristics. The carboxylation is likely to be between 3 and 6 per cent. At the highest level it causes excessive stiffening in the presence of zinc oxide and coagulating cations, raising the modulus to such an extent that the finished latex film becomes paper-like to the touch. Exceptional physical properties, as well as enhanced solvent resistance, can be achieved by this degree of polymer modification.

The degree of sulphur crosslink density following vulcanisation also affects the finished product performance. Using sulphur levels in excess of $1-1\frac{1}{2}$ parts will not give the best physical properties, but solvent resistance is certainly enhanced when the sulphur level reaches 2–3 parts.

Nitrile latex is used for coagulant-dipped supported and fabric-lined glove manufacture. It is more difficult to manufacture gloves from nitrile latex than natural rubber for the following reasons:

(i) The coagulated polymer films have a lower wet gel strength.

(ii) The polymer is less permeable, which is important in sequential dipping operations and processing.

(iii) The low wet gel strength does not permit the same degree of prevulcanisation possible with natural rubber, although for satisfactory dipping some prevulcanisation or maturation is essential for consistent dipping performance.

(iv) Aqueous coagulants function satisfactorily for nitrile glove manufacturing operations but an alcoholic coagulant is preferable as this enhances the inherent wet gel strength of the polymer.

(v) The low total solids required in the dipping bath for satisfactory coagulant penetration, permitting sequential latex dipping, good leaching, satisfactory drying and vulcanisation, is reflected in the excessive shrinkage of the gel during product manufacture. The degree of carboxylation, together with the level of zinc oxide, contributes significantly to the rate of latex film deposit and the degree of syneresis.

(vi) The production line cycle time for a similar product made from nitrile latex as opposed to natural rubber latex is increased by 50 to 100 per cent.

(vii) Because of low wet gel strength, former design is a critical feature of nitrile glove dipping if a transfer roughened article is being manufactured. Pattern designs must be minimised in depth to avoid the coagulated latex film from cracking on the edges of the roughened pattern.

(viii) Adequate ventilation and extraction facilities are essential on manufacturing plant using nitrile latices as the free monomer level for acrylonitrile must not exceed its low threshold limit. By judicious plant design, and careful monitoring, safe working conditions can be established.

A typical formulation with nitrile latex is given in Table 9.2.

Polychloroprene latices

Polychloroprene is similar to natural rubber in gel strength. Many of the comments made relating to the manufacture of dipped goods using nitrile latex apply to polychloroprene, but it is worth noting specifically these points:

(i) The small particle size of the polymer and the inherent polymer property of low permeability makes this latex more difficult to dry and vulcanise.

(ii) The polymer structure does not lend itself to rapid vulcanisation at the low temperatures required for latex dipped goods. For satisfactory vulcanisation, the latex film has first to be dried and then cured at at least 140°C. Consequently manufacturing plant producing polychloroprene dipped articles will have double or triple the production cycle compared to that of natural rubber.

(iii) It is preferable to use an alcoholic coagulant to ensure satisfactory coagulant penetration and latex film build-up, because of the sealing effect against a purely aqueous coagulant.

(iv) Although dipping polychloroprene latices are non-carboxylated, it is sound compounding practice to include 5–10 phr of zinc oxide in the

TABLE 9.2
Typical nitrile latex formulation

Formulation	Houseware glove compound	
	Dry	Wet
40 per cent nitrile latex	100	250
50 per cent ammonia 0·88	0·50	3·20
50 per cent sulphur dispersion	1·50	3·00
50 per cent zinc oxide dispersion	2·50	5·00
50 per cent zinc diethyl dithiocarbamate dispersion	2·00	4·00
50 per cent titanium dioxide dispersion	2·00	4·00
pigment dispersion	as required	
10 per cent ammoniated casein dispersion	0·015	0·15
10 per cent maleic anhydride–styrene copolymer carboxylated resin solution	0·50	5·00

N.B. The ammoniated casein is a thickener and the carboxylated resin is functioning as a latex stabiliser.

formulation. This will maximise physical property development and rate of cure, as it acts as a hydrochloric acid acceptor.

(v) Storage of the latex will result in a progressive lowering of its pH, which has to be kept in the 10·5–11·0 region by adjusting with dilute potassium hydroxide solution.

(vi) The polychloroprene latex film will form a bead satisfactorily when it is adequately dry.

(vii) The finished product is generally of a lower strength than natural rubber but has a similar modulus characteristic and excellent oil and fat resistance. Its chemical, light, and ozone resistance is exceptional.

Typical formulations with polychloroprene latex are given in Table 9.3.

THE DIPPING CHARACTERISTICS OF NITRILE AND POLYCHLOROPRENE LATICES
For successful production dipping of these polymers a controlled balance of colloidal properties must be achieved. The coagulant must not set the latex film too rapidly, as this will seal the inner surface of the coagulated latex film. Adequate stabilisation to slow this setting rate, without reducing gel strength, is the aim. This can be achieved by stabilisers such as casein and surfactants and careful control of the latex pH.

TABLE 9.3
Typical polychloroprene formulations

Formulations	Houseware glove no. 1 dip		Balloons	
	Dry	Wet	Dry	Wet
50 per cent polychloroprene latex	100	200	100	200
33·3 per cent sodium sulphated methyl oleate solution	1·0	3·0	1·0	3·0
10 per cent sodium lauryl sulphate solution	0·1	1·0	0·1	1·0
50 per cent zinc oxide dispersion	7·5	15·0	7·5	15·0
50 per cent diphenyl guanidine dispersion	1·0	2·0	0·5	1·0
50 per cent thiocarbanilide dispersion	1·0	2·0	0·5	1·0
50 per cent sulphur dispersion	1·0	2·0	—	—
50 per cent antioxidant dispersion	—	—	2·0	4·0
10 per cent potassium caseinate dispersion	0·025	0·25	0·025	0·25

N.B. Antioxidant is not usually justified for the expected life of the product as the inherent oxidation and ozone resistance of the polymer is so good.

COMPOUNDED LATEX TESTING

Without adequate testing procedures it is difficult to successfully control the latex dipping process. Some of the testing methods leave much to be desired in terms of correlating an observed effect with a specific cause. Under these circumstances it is best to maintain a system of compounded latex testing which serves as a historical monitor against manufactured product quality.

It is unfortunate that the most sensitive testing methods available do not evaluate the performance of a latex dipping compound more effectively than the production plant itself. However, if the following tests are employed, to accumulate data on a regular basis, appropriate action can be taken if the compound deviates from the norm.

(i) Latex compound viscosity at 25°C versus the total solids of the latex bath. These tests indicate overall colloidal stability of the compound. Viscosity is determined by a flow cup or Brookfield viscometer.

(ii) The heat sensitivity of the compound can be evaluated by measuring the increasing viscosity of it whilst being held for an extended period at an evaluated temperature of 35–40°C, depending on the formulation under investigation.

(iii) Regular pH determination is necessary because of its bearing on general colloidal stability, and is reflected in the straight and coagulant dipping behaviour of the latex.

(iv) The International Standard Test for coagulum in rubber latex can be applied to the matured compound successfully and indicates the level and trends in terms of particle agglomeration when the stabilised diluted latex is sieved through 300–400 mesh.

(v) The degree of compound maturity, or the extent of compound prevulcanisation, must be monitored on a regular basis so that an historical picture of the process can be established with respect to compound performance as a dipping latex and as a finished product.

The most effective means of establishing the extent of latex cure is by solvent swelling of very thin latex film. These films are air-dried and swollen in a suitable solvent for a pre-determined period.

The well established chloroform test,[3] where equal volumes of natural rubber latex and chloroform are stirred together until they coagulate as a cohesive lump, is an extremely subjective one. The degree of cure is related progressively as the cohesive lump becomes less and less tacky, ultimately forming a particulate crumb.

MANUFACTURING CYCLES FOR PROTECTIVES

The basic layout for a manufacturing plant to produce protectives (contraceptive sheaths) is outlined in Fig. 9.1. There are two basic types of plant that are in current use.

  (i) A low output, higher manufacturing tolerance type.
  (ii) A high output, lower manufacturing tolerance type.

The choice of plant selected by individual companies will obviously depend upon their level of expertise in latex technology, and the commercial consideration of the market to which products are being supplied. It is less economic to manufacture with a total production cycle time of 20 min as opposed to a more rapid output over a total cycle time of 10 min. There are various aspects of each process which must be considered and which are essential for successful protectives manufacture.

The low output manufacturing plant

It is very unlikely that this machine will be of a batch dipping configuration, and will almost certainly be a continuous chain dipping operation. The formers will be stainless steel or glass, and mounted at former centres to give adequate clearance between them. This is important if the plant has sharp curves to negotiate. The drying and vulcanising sections of the installation are heated by:

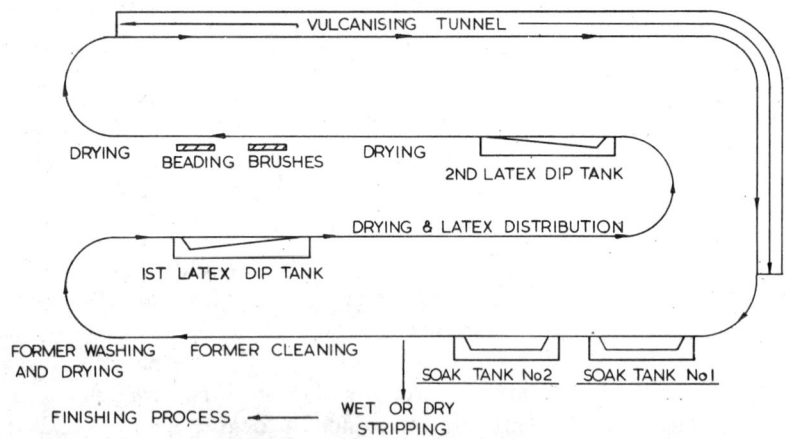

FIG. 9.1. Continuous manufacturing cycle — protectives.

(i) Hot air from direct or indirect gas firing or another indirect source, e.g. by heat exchanger using electricity, steam or hot oil;
(ii) By infra-red heating panels;
(iii) A combination of both methods above.

In the drying stages of the operation a drying airflow can be maintained to give an improved drying rate, although this is not critical on a low output machine where there is adequate tolerance. Climatic conditions can have a significant effect on this phase of the production.

Former cleaning

As with all the stages that will be described, each has its own critical aspects. This is equally true of former cleaning.

The general practice is to use non-volatile cleaning chemicals so as to avoid corrosion of the moving mechanical parts, and to confine the material to the dipping portion of the former or mould. Typical cleaning materials are inorganic alkalis or acids. These chemicals are combined with a scrubbing action to remove deposits formed during the protectives manufacture.

It is extremely important that, following the application of cleaning chemicals to the formers, the chemicals are thoroughly removed with copious washing by warm, soft water, and then dried by passing the formers through a heated tunnel. It is unacceptable to use hard water which would deposit traces of inorganic salts on the former leaving sites for potential hole formation in the latex film.

The latex tanks and dipping technique

Protectives manufacture usually employs the straight dipping technique, i.e. no chemical coagulation is used. In addition, two latex dips are used to minimise the possibility of pinhole imperfections in the film.

The dipping tank construction is likely to be of glass, reinforced resin, or polished stainless steel. There are three basic designs in common use (Fig. 9.2).

(i) The simplest (A) is a narrow, rectangular tank. During dipping the latex is forced to one end of the trough, which needs to be sufficiently wide for the latex to flow back to the entry end of the tank without causing too much turbulence between the return flow and that of the formers going in the opposite direction. A net result of this is that the trough tends to be large, holding a significant quantity of latex which exceeds the preferred minimum.

(ii) This simple trough can be improved by the introduction of a pipe connecting the exit end of the tank with the entry end (B), which reduces the

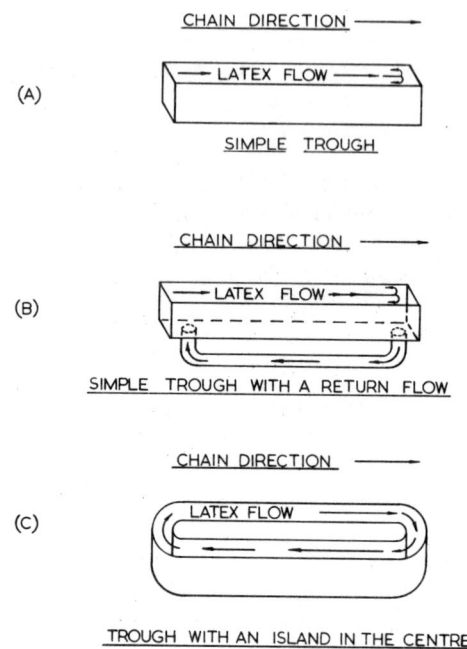

Fig. 9.2. Protectives dipping tank designs.

build-up of latex at the exit end. Latex turbulence is minimised by returning a significant proportion of it down the pipe to the entry end, by the pumping action of the formers which causes a higher level of latex at the exit end or by the introduction of a pump in the return pipe.

(iii) Design (C) is a more sophisticated tank where the formers dip on one side of a central elongated island, and the driving action of the formers passing through the latex drives it around the island and down the outside channel back to the former entry point. It is critical in this design that, when minimising latex tank volume, sufficient clearance between the tank walls and the island is allowed, so as not to create excessive fluid drag and turbulence.

General design considerations should also be taken into account. If the ends of the tanks are not rounded, so that maximum flow along the channel edges is not achieved, then creaming will result. Cream will break off into the latex and cause defective dips. The tanks should be refrigerated to maintain a constant temperature of latex during manufacture. The ideal situation would be where the refrigerant, which surrounds the latex bath, is maintained at minimum temperature differential from that of the latex. It is unwise to reduce the coolant temperature to below 10°C, or to allow the latex temperature to rise above 30°C.

Before covering the actual dipping technique, a few preliminary comments are necessary concerning the dipping compound. Whether using a pre-vulcanised or unprevulcanised latex, certain basic features are essential for successful straight dipping; these are indicated below.

(1) The compound must be mature, i.e. unlikely to change significantly during production.

(2) It will be of low viscosity with adequate stability with some thixotropic character. This allows it to be deposited on the former surface, yet not drain excessively during withdrawal from the latex.

(3) A measure of heat sensitivity must be designed into the latex compound to enable it to set rapidly during the subsequent drying process. This is controlled by the zinc ammine complex formation, which in turn is controlled and modified by the introduction of fixed alkali. The greater the concentration of fixed alkali the less heat sensitive the compound becomes.

THE DIPPING TECHNIQUE AND DRYING PROCESS

In principle it is desirable to enter the latex as fast as possible, and to withdraw from the latex again at maximum speed. These speeds depend upon the latex compound and the dipping tank. With a slower output plant, the maximum entry speed is that which does not cause entrainment of air, and the withdrawal rate is controlled by the viscosity and colloidal stability of the latex. A typical withdrawal rate is within the range of 1·25 to 1·50 s per centimetre of vertical movement. It is usual to have a variable withdrawal rate which is faster over the top part of the product, in order to pick up more latex deposit in this area, which compensates for subsequent latex drainage. The ultimate aim is a dip with as even a wall thickness as possible.

As soon as the former has left the latex after the first dip, it is essential that it is raised to the horizontal, or above, to minimise downward latex flow, and to prevent the formation of a blob at the tip. This will be difficult to achieve if the hangdown time in the vertical attitude exceeds 2 or 3 s.

The main concern, having left the latex, is to achieve an even distribution of the latex deposit, and to this end it must be set as soon as possible by heat. During this critical period its attitude will be varied from the vertical by 45° to 135° with continual rotation. The maximum heat input must be applied immediately after the first dip, followed by progressive temperature reduction so that the former has an opportunity to cool to its lowest temperature prior to entering the second dip. If the former is too hot, heat sensitisation of the latex in the second dip will occur causing destabilisation.

If two dips are made in the same latex compound, the pick-up of latex deposit from the first dip on the former will be greater than the second dip made over the dried, first dip, latex film. Consequently, the second latex dip tends to be more fluid than the first, and is therefore sometimes made less stable.

The entry and exit from this second dip, and its subsequent distribution attitudes in the drying section, may be the same or different to the number 1 latex dip and drying tunnel. Drying out in the bead or ring-forming section of the product should result in a film that is apparently dry, but not dried to such an extent that it is difficult to roll it down the former with the beading brushes.

The general principles of high heat input at the commencement of the second dip drying, in the non-beading area of the product, should be observed. This can then be followed by as much heat as can be effectively applied without causing blistering of the latex film.

The beading operation

In order to prevent the bead or ring from forming a figure eight configuration when the product is removed from the former, it is important that the soft nylon bristle brush, which progressively moves down the formers as they pass through this section, is applied in two phases. The first half of the bead roll-down is with the former rotating in its natural direction; and for the second half the rotation of the former is reversed by a positive mechanical drive.

Vulcanisation

Vulcanisation is achieved by a progressive temperature increase along the oven length. The gradient on a slow output plant will be in the range from 80 to 120°C. If ultra-fast accelerator systems are used, then it is possible that the maximum temperature will not exceed 100°C. This is desirable, as it reduces the chance that residual moisture in the product will be trapped between the product and the former, causing steam blistering.

Leaching and stripping

The leaching process will depend upon the method of stripping.

Dry Stripping

After vulcanisation and prior to removal, the former with the product on, drops into a leaching tank containing soft or deionised water maintained at a temperature of up to 100°C. After leaching, to remove water-soluble compounding ingredients, the former passes through compressed air jets to remove water from the product surface. A dusting powder is applied to the rubber surface to eliminate its tack.

One of the main problems associated with the application of powder is that it is very difficult to contain, and prevent from getting on to the moving parts of the chain. Only when the outside of the protective is rendered tack free can it be stripped successfully by the brushes.

The dry stripping technique involves the use of two angled beading brushes mounted on either side of the travelling formers. As the formers pass through the stripping brushes the product rolls up on itself. An additional brush may be required to remove the product from the former tip.

It is generally recognised that only high modulus products strip satisfactorily using this technique. When the stripping brushes roll the product down, it momentarily exposes an undusted inner tacky surface prior to it being coated by the outside surface as they roll together. This can give rise to internal surface creasing.

The advantage of the dry stripping technique is that the product is ready to be tested and graded without being involved in a wet application of powder to render it tack free, and the subsequent drying step that would then be required (see 'Wet Stripping').

Wet Stripping

This method of stripping is particularly advantageous when manufacturing low modulus or ultra thin protectives. It has the merit that no powder application is required on the manufacturing plant facility, thus maintaining a dust free environment.

As the technique of wet stripping implies, the finished product has got to the state where it can be washed off the former by using water jets. For this to be achieved satisfactorily, the latex compound itself must permit the penetration of near boiling soft water in the leach tank, to enable it to penetrate the latex film and reduce its adhesion to the former. Following this treatment it should only loosely adhere so that it can be flushed off the former as it passes through water jets. The wet products, collected in batches, are then washed in an anti-tack medium. When they are tack free they can be tumble dried using minimum heat to avoid unnecessary oxidation of the very thin product film.

Solvent Stripping

This technique is a practical but little used method for removing the product from the former. The procedure is to dip the protective, on the former, into a solvent that will swell the rubber off the former without making it tacky, and then blowing the swollen product off the mandrel with a pneumatic jet into an anti-tack dispersion. The products are then tumbled in the dispersion and dried as in the wet stripping method.

The high output manufacturing plant

The essential difference between a high output machine and a low output one is in the reduction of manufacturing tolerance that occurs with the decrease in production cycle time. In order to run these plants efficiently, very close control must be maintained over every step of the process, from the preparation of the latex compound through to the ultimate product manufacture. It is obviously an advantage if automatic control or information feed-out of temperatures, etc., is provided for the plant controller. This makes the job of maintaining satisfactory operating conditions much easier. The machine itself will be very similar to the one previously described except that for each manufacturing stage there will be less time available to complete it. This precludes the use of unprevulcanised latex, as the cycle time will be too short to permit satisfactory drying and vulcanisation.

THE DIPPING TECHNIQUE

It is still possible to dip into the latex while maintaining a vertical attitude on the former, but it is more likely that the former will be attached to the chain in such a way that it can pivot about the chain and swing down into the latex. This enables the former to be immersed and withdrawn in the shortest possible time. The latex troughs will be longer and therefore hold much more latex. A careful control of the state of prevulcanisation of the latex is even more necessary, otherwise variable vulcanisation and stripping will occur.

THE DRYING AND VULCANISATION PROCESS

The general principles described on a low output plant are applicable. With a fast running plant there is an advantage that the former attitude can be adjusted more rapidly, allowing more even film distribution.

The most efficient drying and vulcanising processes involve the judicious use of blown air. The air must be clean, and the flow rate must be balanced by the heat input, which is an economic consideration.

General observations
In the ideal protectives manufacturing facility the following features would be apparent.

(i) The manufacturing area would be dust free. It would not be economically justified, although an ideal situation, to feed the manufacturing area with filtered air. A clean room environment would be an unwarranted and expensive degree of technical sophistication.

(ii) The temperature in the manufacturing plant should preferably be kept constant. Practically it is acceptable to have a $\pm 5°C$ variation, providing that the plant is either self-adjusting or controlled by a vigilant operative to compensate for temperature differences that might occur.

(iii) Draughts which could give localised temperature variations must be eliminated or minimised.

(iv) All powdering processes, used in the finishing operation on products after stripping, should be housed in a separate facility away from the dipping plant.

UNSUPPORTED GLOVE MANUFACTURE

The basic processes for surgeons and houseware glove manufacture are similar in general outline.

The possibility of using unprevulcanised or insufficiently matured latex compound is very unlikely, because the thickness of the latex deposit will prevent rapid evaporation of water, and slow the subsequent vulcanisation to an unacceptable rate. It is preferable to use a totally prevulcanised latex, or a blend of unprevulcanised and prevulcanised compound.

The use of transfer roughened glove dipping is almost universally applied these days, because to apply roughening to a latex film by a chemical process dip is costly, hazardous and more difficult to control.

Transfer roughened moulds form the patterning on the inside of the glove while it is on the former. Chemical roughening is applied after the last latex dip, and consequently the glove has to be reversed after stripping.

Surgeons glove manufacture
These days the standard requirements for a surgeons glove manufactur-

ing facility are high, and are expected ultimately to become quite stringent. Whilst a clean air standard for medical products is generally required, a reasonable result is achieved if:

(i) A high level of maintenance of the plant is ensured;
(ii) Good housekeeping is observed;
(iii) Operatives touching the finished products are wearing gloves and are suitably clothed to prevent contamination of the products;
(iv) Surgeons gloves passing through any wet or dry dusting process are maintained in a bacteria-free condition, e.g. by the use of bactericides in wet systems.

The former design

The general principles have been discussed earlier in the chapter, but it is important to point out specific aspects relative to surgeons gloves.

Surgeons gloves can be roughened or smoothed over the finger and hand area. In the case of neuro-surgery or ophthalmic surgery, very thin gloves are preferred, whereas with the less delicate surgical operations it is often advantageous to have a degree of roughening on the gloves to improve grip.

These varying requirements do have a significant effect on the manufacturing procedure or latex compound formulations, because the level of roughening depth influences the overall coagulant and latex pick-up. This will be greatest if the roughening is high. The drying of the coagulant and the gelling of the latex will also be affected.

The thinnest gloves can obviously only be manufactured on smooth formers.

Former cleaning

The fundamental principle of keeping the formers clean during surgeons glove manufacture is a sound one. A coagulant dipping technique results in the coagulation of a thin latex film deposit, susceptible to film imperfections if it were not deposited on a clean surface.

Whether the process is continuous, or by batch dipping, the favoured procedure is to pass the formers through a clean bath of alkali or acid cleaning material on every single production cycle. The formers must be thoroughly washed free of the cleaning material and scrubbed by brushes in areas of the former where dirt deposits are particularly likely to accumulate.

Former pre-heat tank or oven

It is quite normal to use a former heating section prior to dipping in the coagulant, although residual heat in the mould is often adequate. The most efficient way of pre-heating is by immersion in hot water, preferably soft water, which is free from scum or dirt. It is essential that the formers are not contaminated prior to the coagulant dip, as this will give rise to an imperfect latex film deposit.

A hot air oven is not as rapid in equalising the temperature as water, but it is quite effective given sufficient time.

The temperature of the precoagulant former heating bath should ensure a former temperature of 5–20°C below that of the coagulant prior to dipping into it.

Coagulant bath

The Aqueous Coagulant

The advantages of an aqueous coagulant are the economy and safety of using a non-solvent system. A typical formulation is indicated below:

	Party by mass
Hydrated calcium nitrate	20
Parting aid (inert powder)	1–2
Surfactant	0·05–0·25
Tap water	up to 100

Despite the high temperature of an aqueous coagulant, 70–80°C, its slow evaporation rate with time allows too much drainage from the former as it is withdrawn from the bath. The resulting coagulant concentration differential is reflected in the latex film thickness which is thinner at the top of the product.

The coagulant flow is minimised by the introduction of an inert parting aid, which is usually a powder. It is effective because, as the water evaporates, the powder increases the viscosity more rapidly and at the same time functions as a surface roughening reservoir slowing down flow. The parting aid also minimises latex film tack to the former and aids the stripping operation.

The surfactant in the coagulant is usually an anionic or non-ionic water-soluble type. Its primary functions are to fully disperse the parting aid powder in the coagulant, and to ensure that the coagulant itself is an effective wetter of the former.

The Alcoholic Coagulant

While an alcoholic coagulant has the disadvantage of being less economic and potentially a fire risk, its flow and evaporation properties are such as to warrant its continued use in the manufacture of surgeons gloves. A typical formulation is detailed below:

	Parts by mass
Hydrated calcium nitrate	15
Parting aid (inert powder)	1–2
Cationic or non-ionic surfactant	0·1–0·2
Tap water	10
Industrial alcohol	up to 100

Alcoholic coagulants are generally run at a temperature of 30–40°C. Coagulant drainage can be minimised because the evaporation rate is very rapid. This, combined with a variable withdrawal speed, which is more rapid at the commencement of former ascent, slowing down in a uniform fashion until the fingers leave the coagulant surface, attains an even deposit of coagulant. Typical withdrawal speeds would be 6 s per 10 cm over the cuff area, slowing to 15 s per 10 cm over the fingers. It is beneficial to change the attitude of the former when it leaves the coagulant, to distribute the concentration of coagulant at the finger tips more evenly over the fingers. Alcoholic solutions evaporate so quickly that they are not always inverted in this way.

Alcoholic coagulants have the ability to cover uniformly uneven former surfaces, e.g. finger crotches or roughening patterns. This is particularly important if the coagulant strength is low because a thin deposit of latex film is ultimately required.

The latex dip tank

As a relatively thin coagulant dipped film is normally required for surgeons glove manufacture, only one latex dip is applied to the former on a weak coagulant strength deposit. The degree of dryness must not exceed the point where the coagulant begins to crystallise on the former, otherwise the subsequent latex deposit will trap air in the crystallised coagulant coating causing an imperfection in the latex film. Excessive wetness of the coagulant will also dilute the latex, weakening the gel as it is coagulated. It will wash off the former in the general bath agitation as the former enters the latex.

The entry speed at which the former strikes the latex is critical. An initial entry speed of 6 s per 10 cm is recommended. Once the former passes the thumb crotch, the entry speed can be increased without any detrimental effect. The total immersion time of the dipped former length will usually be quite short, e.g. 5–20 s. The former is elevated from the latex at a graduated withdrawal speed, commencing at 6 s per 10 cm and slowing down to 15 s per 10 cm over the fingers.

At the commencement of full immersion, the former dip length is increased by about 1 for the first second or two of the total dwell time. This gives a thin coagulated deposit of latex which can be rolled down, after dipping and drying, to start the ring or bead formation at the top of the gauntlet of the glove. It is unwise to dip the former above the coagulant to produce a thin latex film deposit, because it will adhere too strongly to the former to enable the bead to be rolled up.

After withdrawal from the latex, the former can be moved in an attitude to minimise latex blob formation on the fingertips, and to prevent general downward drainage into that area.

Drying and beading section

Before the bead can be formed, the latex deposit in the top cuff area must be sufficiently dried to become tacky and yet to have sufficient gel strength to permit beading either manually or by mechanical means. It is advantageous to rotate the formers if the beading method is mechanical. Small rotating brushes roll the rubber progressively down the former as the former itself rotates.

Additional drying is necessary if it is intended to brand the size onto the cuff. The latex gel must be sufficiently dehydrated to be able to withstand the physical impact of the rubber-based ink on the branding head, striking the gel surface either manually or automatically.

The leaching process

This section is particularly important in surgeons glove manufacture because of the critical end use of the product. The aim of the leaching process is to remove all water-soluble ingredients to an acceptably low level, e.g. calcium salts and other water-soluble non-rubber materials that are in the coagulated film. If this is not done, skin irritation can result when the product is in use.

A satisfactory leaching time, using soft water, would be in the region of 10 min at a temperature of between 70 and 80°C. The water-extractable materials must not exceed the maximum permissible limit expected in the finished glove. This would be typically 0·5 per cent approximately.

Vulcanising and drying

For a prevulcanised latex, 30 min should be quite adequate to achieve a satisfactory drying and cure. This time would have to be doubled for an unprevulcanised latex. An expected temperature gradient from 80 to 120°C would be quite normal, but would be dictated by the rate of loss of water from the film, and the bead in particular. If in the first stage of the drying section temperatures are too high, then the bead will inflate. Equally if the product becomes too dried out, it adheres excessively to the former and is extremely difficult to remove at the stripping point.

The stripping point

In the more sophisticated manufacturing units, automatic stripping is used, but more generally this is done manually. The product must be relatively free on the former, and is usually removed by pulling the bead down towards the finger tips until the product is turned inside out. It will not be possible to do this successfully unless a lubricating fluid or powder is applied continuously to the inside and outside surface during stripping. Normally a slurry of water and sterilisable starch powder is used. The wet products are dried in heated tumblers at a temperature of 60–70°C for a time determined by the batch size and air flow through the tumbler.

Following drying, the products are kept in light impervious containers while they await grading and packing, which is carried out in a dust-free, purpose-built area. The packed gloves are then sterilised ready for use, by ethylene oxide or the preferable irradiation procedure.

Houseware glove manufacture

Many of the facets of houseware glove manufacturing are similar to those of surgeons gloves. A thin single dip houseware glove without flock adhesive lining is virtually the same. Figure 9.3 is a pictorial representation of a houseware glove dipping plant.

The essential differences between houseware and surgeons glove manufacture is outlined below.

The coagulant bath

This will invariably be an aqueous coagulant and the strength will be in the range of 20–60 parts by mass of hydrated calcium nitrate.

The first latex dip

The dwell time for this dip can vary typically from 15 to 60 s. The long immersion time is necessary when a much greater deposit is required.

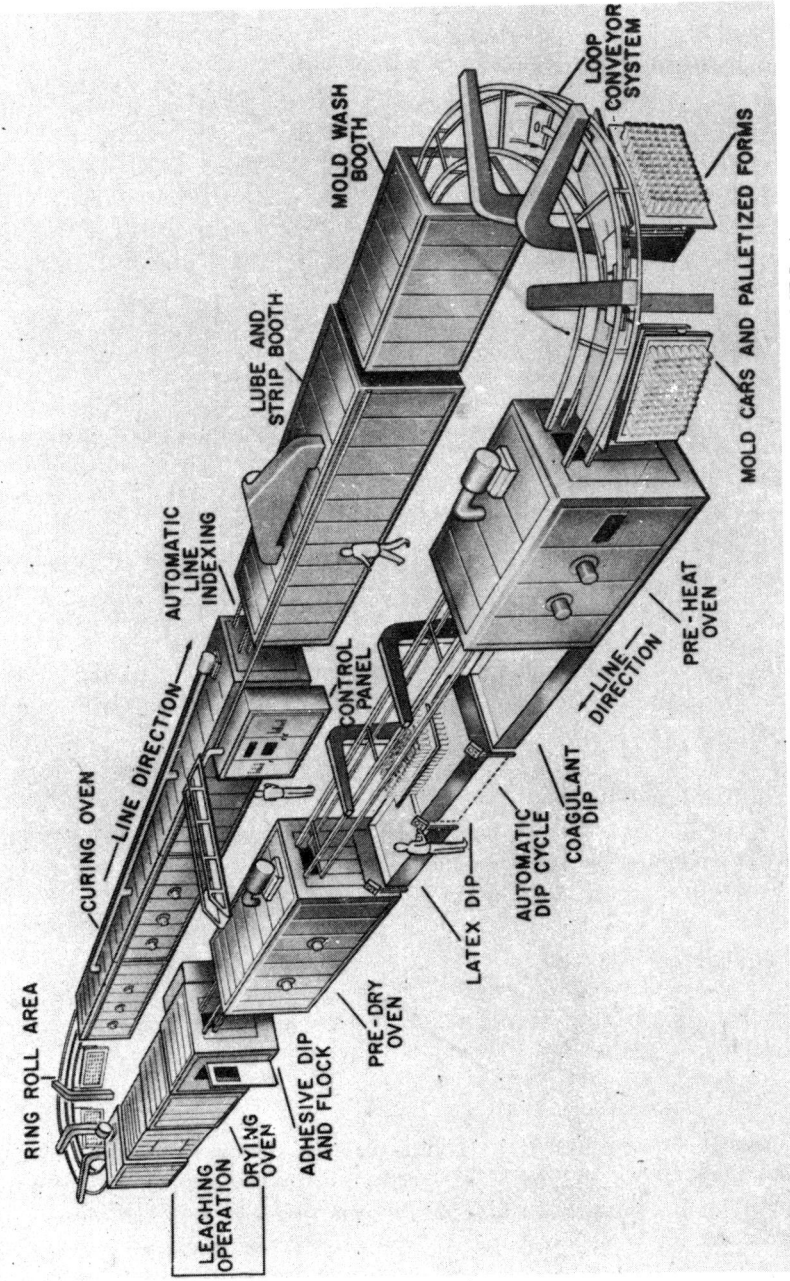

FIG. 9.3. Pictorial representation of a houseware glove dipping plant. (*Courtesy:* LME Inc.)

Drying and Beading Section
This is similar to surgeons glove manufacture.

The Second Latex Dip Tank
This facility is necessary for extra thickness, or cotton flock application. It will only be effective if the coagulant can penetrate the first dip satisfactorily. The hydrostatic pressure during immersion in the second dip consolidates the gel, and immersion time will obviously play an important part.

If it is intended to spray cotton flock fibres onto the second latex deposit, the following conditions must be met:

(i) The stability of the latex must be high to resist coagulation;
(ii) The rubber content should be more dilute than the first dip and it should be modified with hydrocolloid thickeners to reduce uncoagulated latex flow on the surface when withdrawn from the latex;
(iii) The second dip should have a thin uncoagulated surface layer and a fully coagulated layer on the inside which adheres to the first dip.

If the latex is too stable, the flock spray will be totally wetted by the latex and sink into it and not give the desired soft fibre finish at the surface.

Drying and Leaching
After flock application, drying for a few minutes at 100°C is necessary prior to leaching. Adequate leaching is achieved for 10 min in soft water, at 70–80°C.

Vulcanising Tunnel
The drying and vulcanising section is similar to surgeons gloves, but will have the temperature gradient adjusted to allow for the thicker film deposit.

Stripping and Drying
Following vulcanisation the product can be stripped from the former using a wet or dry method. A dry method is obviously preferable as only a minimal tumbling in heat is required to dry the products out completely. An anti-tack finish can be applied at this stage, e.g. talc.

Chlorination

This process is frequently applied to houseware gloves, giving a slick anti-tack finish to the outside of the product, as well as giving it increased surface resistance to detergents and fats.

The products are tumbled in chlorine water at a concentration of 800 to 1 200 mg/kg for 15 to 30 min, after which the chlorine is neutralised and the gloves are washed and dried. The products can then be graded and packed.

Following chlorination, the slick finish on the inside of a non-flocked product allows an 'easy on, easy off' advantage for the consumer.

ELECTRICIANS GLOVE MANUFACTURE

These products are manufactured by conventional coagulant dipping techniques, although the properties required for the finished product are very specific to their end use.

Electricians gloves are much thicker than the normal housewear gloves, and need multiple dipping, successively, in coagulant and latex, to achieve the required thickness. One of the problems that could be encountered relates to delamination of each latex dip from the preceding one. This is avoided by the judicious control of drying temperatures and times, and the use of an alcoholic coagulant without a parting aid. Adequate drying between dips could require as much as 10 to 15 min at 70°C.

The most critical phase is the leaching process, which is ideally carried out in two stages. The first is 30 min at 50 to 70°C in deionised or soft water. The second stage, after the drying and vulcanisation of the product, and following stripping from the former, is a continuous leaching in cold water for 24 h. This extensive leaching is necessary to remove all hydrophilic materials which are potential electrical conductors.

The drying and vulcanising of this thick product is an extended process, with a temperature not exceeding 100°C until the last 15 to 30 min of the vulcanisation cycle. The preliminary drying and vulcanising should commence at 75 to 80°C, rising to just below 100°C for up to 60 min.

SUPPORTED GLOVE MANUFACTURE

Supported glove manufacture is normally achieved by coagulant dipping processes, although some heat-sensitive dipping techniques have also been employed.

For the successful manufacture of fabric-lined gloves, for industrial or heavy duty use, several important factors have emerged:

(1) The cotton liners must fit the former without undue tension, which would stretch open the weave and allow latex penetration.
(2) The cotton liner fabric must be of adequate thread density and thickness to prevent total latex strike-through.
(3) The strength of the coagulant, preferably an alcoholic calcium salt solution, should be sufficient to achieve an adequate latex deposit, and yet not prevent some latex penetration into the fabric to give good rubber/fabric adhesion.
(4) The stability of the latex compound should be adjusted so as not to penetrate right through the fabric liner before coagulating in the cotton thread matrix.
(5) The latex viscosity, together with (3) and (4), also plays a vital part in controlling strike-through into the fabric, i.e. the higher the latex viscosity the less penetration into the cotton glove liner.
(6) When immersing the former and liner into the latex after the coagulant dip, the speed of entry must be not so great as to cause additional hydrostatic pressure, hence forcing the latex through the fabric layer onto the former. A satisfactory entry speed of vertical travel into the latex would be 1 s per cm.

The former is usually of generous proportions in design, with a polished surface finish. If the surface is not slippery, the loading of liners onto the formers manually would be difficult. It is important that the protruding fibres from the liner are flamed to remove them prior to dipping.

The subsequent steps in the manufacturing process are similar to those of unsupported glove dipping.

The most satisfactory type of plant for supported fabric-lined gloves is a batch dipping type. This gives an opportunity for satisfactory loading and stripping of the liners and gloves, which is a manual operation.

BALLOON MANUFACTURE

Novelty and advertising balloons are usually manufactured by a single dip process using an aqueous or alcoholic coagulant. The more complicated shapes require coagulant modifications to enable successful stripping. The parting aid in the coagulant may be as much as 20 per cent by mass of the total coagulant, which in turn brings its own former cleaning problems.

Essentially balloon manufacture is very similar to that of surgeons gloves, with some important differences which are described below.

(1) If large amounts of parting aids are included in the coagulant, then a bead adhesive would have to be applied separately, before the bead could be formed to make it stick to itself.

(2) Whilst the leaching of balloons is not as critical as for surgeons gloves, an efficient process is used because any residual calcium nitrate in the balloon will taste if inflated orally.

(3) Balloon formers are usually closely packed on a manufacturing plant to yield maximum output. This brings its own problems:

> (i) If an anti-tack coating has not been applied to the products prior to going through the drying and vulcanising sections, then they will stick to each other should they inflate.
>
> (ii) The stripping is one of the more important aspects of balloon manufacture, and the technique used depends upon the shape of the product. Simple shapes, which have a similar neck to body diameter, can often be plucked off a former manually. More complicated shapes require water injection down the neck prior to mechanically stripping from the former between two soft rubber pads. Water jets alone can be used for the more simple shapes of balloon.

(4) Following stripping, the balloons are fairly tacky, and require coating in an anti-tack dispersion of talc or zinc stearate prior to drying.

(5) A 'jazzing' effect is obtained on balloons by dipping them, while on the former, into solvent-based ink floating on a water surface. Alternatively the balloons are passed under a stream of different coloured inks.

Balloons are made in varying thicknesses, from the thin modelling types that can be twisted into different shapes, to the thicker types, e.g.

advertising balloons. The important requirements for this type of product are its resistance to the application of solvent-based printing inks and its ability to remain inflated for extended periods.

THE MANUFACTURE OF BOTTLE TEATS AND SOOTHERS

There are two processes which can be used for manufacture. These are by coagulant or heat-sensitised dipping. As the products are thick, it requires approximately three sequential dips in coagulant and latex to achieve the desired result. Careful drying control is necessary to avoid delamination between the dips.

Typically a dwell time in the latex of 60 to 90 s would be required for each latex dip, and a very slow withdrawal, following total immersion, to minimise blob formation on the products tip.

Once all the latex coats have been formed, it is advisable to consolidate the film and carry out a preliminary leaching in hot, soft water at 60°C for 5 to 10 min. It is best to strip the product from the former using water jets or an air gun, and then leaching it for 4 h in cold water.

To remove the product tack, which would result on the dry product, the introduction of a chlorination process or silicone emulsion rinse would have a satisfactory and non-toxic de-tackifying effect.

Assuming that the latex used is a fully prevulcanised one, then the final drying process can be carried out off the former satisfactorily, on pegs, in an air circulating oven for 16–20 h at 70°C. The dimensions of the former must be adjusted to take into account the subsequent shrinkage that will occur in the product during this final drying.

The most elegant production method for manufacture of bottle teats is by heat-sensitised dipping, using stainless steel formers. Fully prevulcanised latex is used, coating the former initially with a thin uncoagulated dip. Following drying of this, and heating the formers to a carefully controlled predetermined temperature, a dip into heat-sensitised latex with only a second or two immersion time is carried out.

For this technique to be successful in commercial teat dipping production, careful control of the former heating and the latex bath temperature is necessary. The bath temperature should not vary more than $\pm 1°C$, and the heat coagulation point of the latex must be maintained at a constant value. The production method then follows the one outlined for coagulant dipping.

The dipped product, if it is unbeaded, is then trimmed to a precise dimension. On the same rotating turntable, a heated platinum wire is plunged through the top of the teat making the hole.

OTHER PRODUCTS

There are many miscellaneous products which can be made by dipping techniques, and they are all variations of the methods already described, although some, like catheter manufacture, are very specialised with a great deal of know-how involved as well as manual manipulation during the manufacturing process.

In the space available it has not been possible to cover, in a detailed fashion, heat-sensitive dipping of industrial gloves, over-shoes, Wellington boots and products of this type, but they are essentially similar to teat manufacture.

PRODUCT TESTING

Many latex dipped products are subject to requirements of national Standards Institute Specifications, and it is anticipated that these will eventually be consolidated in International Standards Organisation Specifications.

Apart from these specifications, there are other aspects which are important to the finished product and application. Some of these are mentioned below.

Cure

(1) Solvent swelling methods, using solvents like toluene or heptane, can be used to establish the extent of crosslink density in the finished product. product.

(2) Relaxed modulus testing at 100 per cent elongation can also be used, but it is essential that smooth regular films are cut from the finished product for the results to be meaningful.

(3) An estimation of free sulphur is a good indicator of whether or not the product is fully vulcanised.

(4) Tensile strength and elongation-at-break, together with modulus data, yields useful information which is consolidated by repeating these tests after heating the product for seven days at 70°C in an air circulated oven.

Solvent and chemical resistance

The behaviour of latex dipped goods in aggressive environments can be assessed by measuring:

(1) The degree of film swelling following standard times of immersion in the solvent/chemical;
(2) If the solvent/chemical is non-volatile, or the deterioration of the strength of the test piece is not affected by the volatility of the solvent or chemical, then tensile strength testing is an effective measure of the products' resistance following prolonged immersions in these materials;
(3) Particularly relevant to chemical resistance is the careful monitoring of the surface of the polymer film following prolonged immersion in the chemical, and recording the deterioration of the surface against a progressive time span.

Light and ozone resistance

These properties are of particular consequence when the finished products are either manufactured, or intended for use, in countries where high temperature, strong sunlight and significant ozone concentrations are prevalent. The best indication of a products performance is established by observing the effect of exposure under these natural conditions. It sometimes proves to be misleading when laboratory conditions are related to those which occur naturally.

FUTURE DEVELOPMENTS

It seems unlikely that new dipping polymer latices are going to be developed in the foreseeable future. The emphasis is almost certainly going to be on making the best use of the materials that are already available. This is likely to be by modifying existing polymer latices, as is already in evidence with natural rubber, or by using blends of similar or dissimilar latices to achieve properties which are not possible to obtain from single polymer use.[4]

Economic pressures will dictate an increasing use of polymer laminates and more efficient manufacturing techniques. The most effective means of rapidly drying latex dipped films could become critically important, if latex dipping is to remain competitive with other manufacturing methods as these become more sophisticated.

REFERENCES

1. BLACKLEY, D. C., *High Polymer Latices*, Vol. 1, Applied Science Publishers Ltd, London, 1966, 35–43.
2. BLACKLEY, D. C., *High Polymer Latices*, Vol. 1, Applied Science Publishers Ltd, London, 1966, 383–91.
3. *Vanderbilt Latex Handbook*, 1954.
4. GORTON, A. D. T., *NR Technol.*, 1979, **10**, 68.

BIBLIOGRAPHY

BLACKLEY, D. C., *High Polymer Latices*, Vols. 1 and 2, Applied Science Publishers Ltd, London, 1966.
KOCH, S. (ed), *Bayer Manual for the Rubber Industry*, 1970.
CARL, J. C., *Neoprene Latex*, Du Pont De Nemours & Co., 1962
Breon Latices Technical Manual No. 5, BP Chemical (UK) Ltd, 1969.
Vanderbilt Latex Handbook, 1954.

Patents
British Patent Number 368,295, *Improvements Relating to a Method of and Apparatus for Making Dipped Rubber Articles*, accepted 1932.
British Patent Number 380,174, *Improvements in or Relating to a Method of and Apparatus for Moulding Hollow Articles of Rubber Latex or Like Siccative Liquid by Dipping*, accepted 1932.
British Patent Number 614,358, *Improvements in or Relating to the Production of Hollow Articles by Dipping*, accepted 1947.
United States Patent Number 2,649,960, *Apparatus for Testing and Sorting Thin Rubber Articles*, accepted 1949.
United States Patent Number 2,649,619, *Apparatus for Manufacturing Dipped Rubber Articles*, accepted 1950.
British Patent Number 1,097,458, *Dip Coating Apparatus*, accepted 1968.

Chapter 10

MOULDED LATEX FOAM

J. C. Fallows
Dunlop Ltd, High Wycombe, UK

Cellular rubbers, produced by the *in situ* generation of gas in a polymer, have been well known for a considerable time, but the production of latex foam, a low density, open cell flexible material, had its beginnings in the 1920s.

Early attempts at making a foam from latex were only partially successful. By introducing stabilising agents and by using mechanical agitation it was relatively easy to achieve a wet froth, but the difficulties in coalescing and coagulating the disperse rubber phase without the simultaneous collapse of the disperse air phase prevented any real progress being made.

In 1929, Dunlop Rubber Company applied for two patents which introduced, *inter alia*, the use of sodium silicofluoride to achieve the controlled gelling of the latex froth, and it was this important breakthrough that was the start of the latex foam industry.

The only successful alternative to the sodium silicofluoride, or Dunlop, process is the that developed by Talalay in 1946 and improved, by the addition of vacuum expansion, in 1955.

Latex foam rubber is used as a cushioning material and its major outlets are therefore in the furniture, bedding and automotive industries. Polyurethane foams, introduced in the 1950s, have had some effect on latex foam consumption but a discussion of this point is outside the scope of this book. Polyurethane foam and latex foam are complementary materials in many ways, have similar end uses and many common methods of test.

PREPARATION OF FOAMS

Latex foam chemistry and technology fall naturally into two disciplines, polymer chemistry and surface chemistry.

Control of the surface chemistry of the latex foam systems is vital to good foam production and the Dunlop process and the Talalay process solve the problems it presents in totally different ways.

The problem is simply stated. It is to generate a stable air/liquid dispersion (a froth) in a rubber/liquid dispersion (a latex) and then to destabilise both systems to give a stable rubber/air system in which both phases are continuous.

Dunlop process

The gelling system employed in the Dunlop latex foam process achieves the required results in a one-stage operation, using zinc oxide and sodium silicofluoride (SSF).

The latex compound contains, in addition to the curing ingredients, ammonia and additional soap. After the froth has been made by mechanical agitation, zinc oxide and SSF dispersions are added as separate additions. The pH of the froth at this stage is in the range 10·0–10·5.

Sodium silicofluoride hydrolyses slowly at this pH as follows:

$$Na_2SiF_6 \rightleftharpoons 2Na^+ + SiF_6''$$

$$SiF_6'' + 4H_2O \rightleftharpoons Si(OH)_4 + 4H^+ + 6F'$$

This hydrolysis brings about gelation of the latex foam in three ways:

(1) The fall of pH due to the presence of hydrofluoric acid;
(2) The adsorptive effect of the silicic acid;
(3) The destabilisation effect of zinc ammines formed from the reaction of zinc oxide with ammonium fluoride.

Very simplistically, by suitable choice of the levels of zinc oxide and SSF and in some cases by the use of secondary gelling agents such as diphenyl guanidine, the reactions leading to the destabilisation of the air/liquid system and the rubber/liquid system can be balanced to give an open cell foam.

Talalay process

In the process developed by Talalay, gelation of the air/liquid and rubber/liquid systems is achieved in separate consecutive steps. The liquid latex froth is frozen, in the mould, at a temperature of $-30°C$. The rapid rise in surface tension causes the air/liquid system to break down and the bubbles to inter-connect, but the increasing viscosity and ultimately solid 'ice-foam' prevent the foam system collapsing. The open cell 'ice-foam' is then irrigated with carbon dioxide gas which reduces the pH and brings about gelation of the rubber/liquid system.

DUNLOP PROCESS

The process for making latex foam by the Dunlop process may be sub-divided into the following operations:

(1) Preparation of compound;
(2) Foaming;
(3) Addition of gelling agents;
(4) Pouring the sensitised foam into a mould;
(5) Foam gelling;
(6) Foam curing;
(7) De-moulding the foam;
(8) Washing and drying.

Preparation of compound

Compounds for latex foam production are based on natural rubber latex, cold polymerised SBR latex or cold polymerised reinforced SBR latex. Blends of latex are frequently used, the ratio in the blend depending on a number of factors, e.g. formulation economics, physical property requirements, filler tolerance. Typical specifications for these latices are as follows:

	Natural latex	SBR latex
Total solids content (per cent)	61·5 min.	62·5 min.
Dry polymer content (per cent of total solids)	98 min.	94 min.
Mechanical stability (s)	540 min.	1000 min.
Volatile fatty acid number	0·2 max.	—
Potassium hydroxide number	1·0 max.	—
Viscosity (mPa.s)	—	400–900
Volatile unsaturates (per cent)	—	0·1 max.
pH	—	10–11
Surface tension (mN/m)	—	32–35

The latex required to make a batch of compound, commonly about 5 tonnes wet, is accurately weighed into a jacketed mixing tank and to it are added the weighed compounding ingredients, with constant stirring to ensure good mixing.

The gelling process is dependent on pH reduction and it is desirable therefore that the ammonia concentration of the latex is adjusted to a controlled level, usually 0·2–0·25 per cent. Where a high-ammonia natural latex is used, it is necessary to reduce the ammonia to this level prior to compound preparation. This is achieved by stirring the latex with an air stream over the surface. Chemical reaction with, for example, formaldehyde, can also be used but as it is expensive and the resulting hexamethylene tetramine can cause variability in the final product properties, it is not a popular method.

For compounds based on SBR latex or blends with a high SBR/NR ratio, it may be necessary to add ammonia to increase the concentration to the desired level.

All solid compounding ingredients are added as dispersions, generally about 40–50 per cent solids content to avoid overdiluting the final compound. It is not proposed to discuss the preparation of these dispersions here except to stress three important properties which the dispersions must have:

(1) the dispersing agents used must be compatible with the surfactants in the latex and in the other dispersions;
(2) The dispersing agents must not interfere with the critical gelling reactions in the subsequent processing stage;
(3) The particle size of the dispersions should be similar to that of the rubber particles in the latex. This is particularly true of the vulcanising ingredients.

Dispersions are generally prepared in ball mills, although attritors have also been used, and the dispersing agents are generally oleates or sulphonates. Frequently, additional stabilisation of the dispersions is achieved by the use of a protective colloid such as powdered glue, or sodium carboxymethyl cellulose.

The compounding ingredients added at this stage are soap (for the subsequent generation of a froth), curing agents (sulphur and accelerators), an antioxidant (to inhibit ageing of the final product) and, optionally, mineral fillers, processing oils and additional stabilisers. Mineral fillers, if they are available in small particle size, may be added directly to the latex as dry powders provided that additional soap has already been added to the latex so that the filler does not destabilise the system by scavenging the stabiliser. If this method is chosen, controlled addition of the filler with efficient stirring is essential to prevent high local concentrations of unprotected filler in the compound.

It is also common practice to make up master batches of ingredients, such as sulphur and accelerators, for ease of handling and to aid quality control.

Typical recipes for compounds based on natural latex and SBR latex are given in Appendix 1 of this chapter.

Foaming
Foaming of the latex compound is achieved by mechanical agitation in the presence of air. The low surface tension and the relatively high viscosity of the latex compound enables a stable froth to be made.

In the early days of latex foam production, batch frothers of the bowl and whisk type were employed and these are still used in low capacity operations. The latex compound is weighed into the bowl of the mixer and the whisk rotated at high speed to raise the foam as rapidly as possible to the required height. As the density of the finished product depends directly on the froth density and therefore the froth height, this is measured accurately. After attainment of the required height, the whisk speed is reduced to comminute the bubbles and refine the froth without further reducing the density.

Modern high capacity latex foam plants use continuous foamers and several commercial machines are available, typical of which is the Oakes Mixer. The latex compound is metered accurately and continuously together with a metered flow of air into the centre of a circular stator fitted with concentric rows of flat teeth. A rotor with similar rows of flat teeth rotates so that the two sets of teeth enmesh with sufficient clearance for the latex/air mixture to flow through to the outer circumference. During its passage from the centre to the circumference, the latex/air mixture is subjected to an intense mixing action resulting in a continuous flow of a uniform and fine cell froth. Rotor speeds are usually between 100 and 400 rpm. The froth density is controlled by means of the latex to air ratio and the output is controlled by means of the latex to air ratio and the output is controlled by the latex pump.

Gelling agents
Zinc oxide and sodium silicofluoride are used as gelling agents and these are added to the froth when the correct density and fineness have been achieved.

In the bowl and whisk method of frothing, a measured amount of zinc oxide dispersion is added to the bowl and stirred into the froth followed by a measured quantity of SSF dispersion. After a further period of blending, the liquid, sensitised, froth is poured from the bowl into a mould.

Needle injection of zinc oxide and SSF is the method employed in the continuous foam operation. Attached to the mixing chamber of the continuous frothing machine is a blender consisting of a barrel containing angled rotating perforated discs which give a low shear blending action. Zinc oxide dispersion and SSF dispersion are injected at a metered rate, depending on the latex throughput, through hypodermic needles inserted in the barrel, where they are continuously blended into the froth. The sensitised froth is conveyed by a plastic hose to the mould. This plastic hose also serves to generate the back pressure which is required for good frothing performance of the continuous foaming machine.

Moulding

The moulds used are usually made from cast aluminium, the lids containing the familiar domed cavity formers. Each mould is conditioned to ensure that its temperature is uniform and constant, generally about 40–50°C, and treated with a mould release agent. There are many alternative mould release agents employed but the most widely used are aqueous solutions of sodium carboxymethyl cellulose or medium molecular weight polyethylene glycols or a mixture of both.

To ensure good demoulding, the mould release solution should be applied evenly and uniformly over all surfaces and allowed to dry. This is obviously best applied by spraying immediately after demoulding a product when the mould is still hot after curing. The mould release solution dries off quickly and serves also as an aid to mould cooling.

In a batch foaming operation, each batch of froth is usually calculated to fill one mould cavity or in some cases a small number of cavities with a small excess. The froth is poured evenly into the mould, and the lid carefully put in place. The lid contains a number of small holes through which air and the excess froth escapes and the weight of the lid is sufficiently great for it to settle on to the flanges of the mould pan, without the need for clamping.

In a continuous operation, a series of moulds are attached to a circular conveyor with lid opening and closing achieved by angled guide rails. In this case the lids are hinged to the mould pans.

The sensitised froth from the mixing machine is fed, from the plastic hose, into the mould where it is uniformly distributed. The final density of the product is dependent on the froth density, which is controlled at the mixing machine, but the foam wastage due to spue is dependent on the volume, and therefore the weight, of the froth placed in the mould. The control of mould filling is achieved with accurate timers, set according to foam throughput, foam density and mould volume. Obviously mould volumes differ and therefore each mould has its own timer.

It will be obvious that some care is required in balancing the froth output with the number and volume of the moulds on the conveyor so that the moulding cycle can be operated continuously. Mould changes can be made, but timers have to be re-set and the complete cycle re-balanced, by adjusting the output of the mixing machine.

Continuous mixers can be programmed to change froth density during the moulding cycle, by changing automatically the latex/air ratio entering the machine. Compound changes can also be programmed into the mixing machine and these are effected by switching the latex compound supply from one feed tank to another. Because of the volume of the equipment and the supply hose and the mixing of the dissimilar froths during their passage through it, the changes are not instantaneous and normally take several seconds to settle down to the new level. This is approximately equivalent to the fill-time for one mould.

Foam gelling

Having poured the foam into the mould and positioned the lid, a short interval is allowed before the mould passes into the curing chamber. During this interval the zinc oxide and SSF additions bring about gelation of the foam. The addition of gelling agents is adjusted so that gelation is complete in about 5 min.

Control of the gelling reactions is vital to good foam production and maintenance of stable conditions is important. The reactions are sensitive to temperature and therefore the latex compound is supplied from jacketed tanks, at a temperature usually of 18–20°C. This temperature rises during frothing and passage through the hose so that the foam temperature at the filling point is 22–24°C. The level of gelling agents and their ratio is important in balancing and controlling the reactions and these are carefully monitored.

Mould temperatures are also important and these are usually controlled to 30–35°C for the pans and 40–45°C for the lids. Mould conditioning is usually effected by a water spray followed by a period in a temperature controlled oven.

Failure to control foam temperature and gelling agent additions can lead to poor foam structure throughout the product. Failure to maintain mould temperatures can lead to defects, such as thick skin or loose skin on the surface of the product.

Curing

Following gelation, the moulds pass into a curing chamber where they are exposed to live steam at atmospheric pressure for 20–30 min. After this time, the foam, although not fully cured, is strong enough to allow the lid to be automatically opened and the product removed from the mould.

Any residual water is removed from the mould and the flanges cleaned of any foam overflow (spue). The internal surfaces are sprayed with mould lubricant and the mould then passes in to the conditioning chamber. Figure 10.1 shows a latex foam cushion being removed from a mould on a continuous moulding conveyor.

Washing and drying

The foam product as it is removed from the mould contains a number of water-soluble, non-rubber constituents such as soap which could lead to poor ageing, poor resilience and bad odour. These contaminants are simply and effectively removed by passing the moulding through a series

Fig. 10.1.

of rollers while clean water is sprayed on to the foam. The excess water at the end of the washing stage is removed by passage through a final squeeze roller. In some operations the washing is carried out batchwise in a large centrifuge. In both cases, the final product contains about 40–50 per cent water which has to be removed by drying.

The modern driers for latex foam products are continuous with rapid air circulation. The drying rate is governed to a large extent by the speed with which water in the product can migrate to the surface, where it is evaporated. High velocity air, which penetrates the foam to a limited extent, accelerates this process.

The wet products are placed on an open mesh conveyor which passes through a chamber with hot air circulating. The air is directed by fans on to the upper and lower surfaces of the product, so that there is a positive pressure on the upper surface to prevent the products being blown about. The air temperatures are generally between 100–110°C and drying times are normally about $1-1\frac{1}{2}$ h.

Drying, *per se*, is normally complete in less than 1 h and the remainder of the time in the drier is to complete the vulcanisation. Mould cure times and accelerator levels in the formulation are adjusted to allow for the post-curing which takes place in the drier.

Drying temperatures should not be allowed to rise to more than 110°C. No increase in drying efficiency will be achieved, because the process is controlled by the rate of migration of water through the product, and excessive temperatures can lead to charring of the dry foam.

TALALAY PROCESS

The process for making latex foam by the Talalay process is similar in some aspects to that described for the Dunlop process and may be divided into the following operations:

(1) Preparation of compound;
(2) Foaming and vacuum expansion;
(3) Moulding the foam;
(4) Foam gelling and curing;
(5) Foam demoulding;
(6) Washing and drying.

Preparation of compound

Compound preparation follows the same lines as described under the Dunlop process. Compounds for the Talalay process are based on SBR latices although blends with a small amount (10–20 per cent) of natural rubber latex are sometimes used. The same general latex specifications are applicable.

Foaming

In the Talalay process as first operated, expansion of the latex compound was achieved by separately mixing into the compound calculated quantities of hydrogen peroxide and a slurry of bakers yeast. The mixture was quickly placed in a specially designed mould where the catalase in the yeast decomposed the hydrogen peroxide to liberate oxygen which expanded the compound into a froth.

The modern Talalay process has dispensed with chemical frothing techniques, which are difficult to operate without very careful control, and has replaced them with a combination of mechanical foaming and vacuum expansion. A high density froth is made on a continuous frother by the same method as described under the Dunlop process, but no gelling agents are added, and the partially expanded froth is placed in a specially designed mould. The lid is closed and a vacuum is applied so that the froth expands to fill the whole mould cavity.

By this elegant method, a range of products of different densities can be produced without the necessity of constantly adjusting the froth density at the mixer. Furthermore, the need to control the froth density with great accuracy is obviated and, as the froth is metered into the moulds by weight and not by volume, control of product density is very accurate.

An important difference between the Dunlop process and the Talalay process is in the mould lubricant used. The complexity of the Talalay mould with its multiplicity of pins would make it a difficult task to apply a mould lubricant over all surfaces. This problem is overcome by the use of an internal lubricant added to the froth just prior to it entering the mould. A quantity of dilute hydrogen peroxide solution is blended into the froth for this purpose, although the exact mechanism of hydrogen peroxide as a mould release is not understood and there are many theories.

Moulding the foam

In the Dunlop process, the moulds are carried on a conveyor past the foaming point and then through a complete cycle of operations. In the Talalay process, the moulds are more complex and much more highly engineered to carry the services, so that they are permanently sited in a hydraulic press.

A diagram showing a cross-section of a Talalay mould is shown in Fig. 10.2. The gelling and curing processes are carried out by heat processes and therefore an important requirement of the mould is to effect temperature changes rapidly and reliably. The mould is fitted with channels through which glycol/water mixtures at precisely controlled temperatures are circulated, and the heat is conducted into and out of the foam by a series of closely spaced pins penetrating the foam from both mould surfaces. Four glycol/water streams are used in the heat exchange system, and their temperatures are normally as follows:

Cold	$-30°C$
Low intermediate	$4°C$
High intermediate	$38°C$
Hot	$110°C$

The mould periphery is fitted with a double groove with a moat between. The outer groove is fitted with a rubber gasket which seals the mould cavity with an airtight fit when the mould is closed. The inner groove is fitted with a replaceable semi-permeable paper gasket through which air or gas can pass but which prevents the passage of the froth.

The mould is prepared by renewing the paper gasket and circulating the low intermediate glycol/water mixture. The partially foamed compound is metered into the mould, an audible signal indicating when

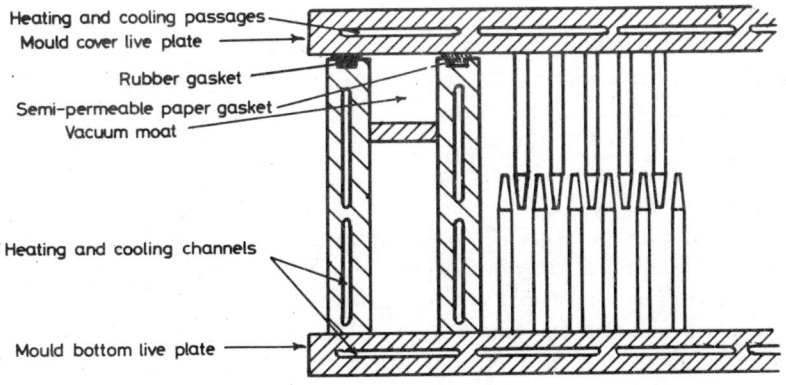

Fig. 10.2.

the correct amount of froth has been delivered, and the hose moved to the next mould.

The automatic mould programmer is then activated which takes the mould through the complete moulding cycle. The mould closes and the vacuum is applied to the moat which withdraws the air from the mould, through the paper gasket. This causes the foam inside to expand to fill the whole mould cavity, the paper gasket preventing the froth passing into the moat.

Foaming and gelling
After the completion of vacuum expansion, automatic valves operate to circulate the cold glycol/water stream through the mould channels to freeze the foam. The rapid rise of surface tension destabilises the air/liquid system and this, together with the growth of ice crystals, causes the air bubbles to interconnect and give an open foam. Normally, the foam would collapse on the destabilisation of the air/liquid system, but freezing the froth prevents this. This process is reversible, and thawing the ice foam at this stage produces the original froth.

With cold glycol still circulating, the vacuum is removed and carbon dioxide gas pumped into the moat where it passes through the paper gasket and the frozen foam. The pH falls from about 12 to 9·25 and the rubber/liquid phase breaks down due to precipitation of zinc soaps, the destabilising effect of zinc ammines and the formation of zinc ammonium carbonate.

Curing
With the rubber system coagulated in a stable foam structure, the foam can now be safely thawed. This is achieved by raising the temperature in stages by successive circulation of low intermediate, high intermediate and hot glycol/water streams through the mould. The final stream raises the temperature to 110°C and the foam is kept at this temperature to effect the cure.

Foam demoulding
At the cure temperatures, the zinc ammonium carbonate formed during gassing breaks down, liberating carbon dioxide and re-generating ammonia. The rise in pH which this causes allows potassium oleate to re-form which aids in removing the product from the mould.

As a further aid in demoulding the foam, valves in the glycol/water streams are operated so that the lid of the mould is hotter than the base.

Because of this temperature difference when the mould opens, the product is withdrawn from the pins in the base and is held in the pins in the lid, where it is easier to strip.

There are a large number of closely spaced pins in the mould and consequently a very high contact area between the foam and the metal surface. At the point of demoulding, the foam is hot and, to prevent damage during this operation, the compound has to be designed to give good hot wet tear strength. Also, the hydrogen peroxide release agent must have been added in sufficient quantity and re-generated potassium oleate and added non-ionic soaps must be present in the foam to lubricate the pin surfaces.

Figure 10.3 shows a mattress block being removed from the mould. Note the closely spaced pins in the mould surfaces to ensure efficient heat transfer during temperature changes.

Washing and drying
The washing and drying of the foam is carried out in the same way as described previously. The presence of pin holes, approximately 6 mm in diameter, in both surfaces of the foam improves the air circulation within the foam and, as the foam thickness between pins is approximately 10 mm, migration of water to the foam surface is enhanced.

Fig. 10.3.

FOAM PROPERTIES

Hardness

The hardness of a latex foam product is the force, in newtons, required to indent the product with a specified indentor, generally 200 mm diameter, to 60 per cent of its original thickness. The method is described in detail in ISO 2439.

To a first approximation, the hardness of latex foam is dependent on density and the relationship takes the form

$$\text{Hardness} = K \,(\text{density})^n$$

where n is a constant, usually between 2·0 and 2·5 which is dependent mainly on the foam structure.

The constant K is dependent on the depth of product being tested, the size of the indentor and the degree of indentation, but if all of these are held constant then K is a measure of the modulus of the continuous polymer phase.

The hardness of a latex foam moulding is also influenced by the geometry of the cores, their spacing and their configuration in relation to each other.

Compression set

This property is a measure of the state of cure of the foam, and is the percentage loss in thickness resulting from holding the foam at 50 per cent compression for 72 h at 70°C and allowing it to recover for 30 min at room temperature. The ISO method is ISO 1856. Figures less than 10 per cent (measured against the original thickness) are generally regarded as acceptable.

Elongation-at-break and tensile strength

These properties are a measure of foam quality and are determined on dumb-bell shaped test pieces stamped out of a foam section 6·35 mm in thickness. Details of the method are given in ISO 1798. Elongation-at-break is largely independent of foam density and is a measure of the quality of the polymer phase. Tensile strength, on the other hand, is not only a measure of the polymer phase but is also, as might be expected, dependent on density.

Fatigue testing

Static fatigue

This test measures the loss in hardness and thickness of a foam sample subjected to a static load for 72 h at room temperature. The load is applied on a circular indentor, the magnitude of the load depending on the initial hardness of the foam sample.

There is no ISO standard for this test, but details are given in BS 3093 and BS 3129. This test is not often favoured now and is rarely specified. It is being replaced by a dynamic fatigue test.

Dynamic fatigue

Several types of dynamic fatigue tests have been proposed to measure the resistance of latex foam to thickness and hardness loss when subjected to cyclic compressions. Early tests were based on 250 000 cycles to a fixed indentation. These tests have now been superseded by a constant load test, details of which are given in ISO 3385.

The cushion is subjected to a force of 750 N for 80 000 cycles and then allowed to recover for 10 min. Changes in hardness and thickness are recorded as well as any physical damage which has occurred.

For any given compound, the hardness losses occurring during constant load pounding are dependent on density, the higher the density the lower the losses. However, the quality of the polymer phase is also important and excessive filler loadings, for example, can give poor results because of physical damage to the foam structure.

In mouldings with large cores, the design and layout of the cores can influence the fatigue behaviour. Cores which are too large or with insufficient wall thickness can give rise to excessive strains during dynamic fatigue possibly leading to physical damage of the foam structure.

Foam economies

Foam cushioning, whether it is for furniture, mattresses, pillows or transport seating is generally supplied to a 'hardness' requirement. In other words, the feel of the foam is specified by quoting a hardness range. On the other hand, the cost of the foam is determined to a large extent by the weight of the product and the cost per unit weight of the compound used. Clearly it is advantageous to use a low cost compound and also a compound and product design that gives a good hardness/weight ratio, always providing that the other properties, such as comfort, fatigue performance, tensile properties, etc., are held at an acceptable level for the particular application of the product.

Product design, i.e. the geometry of the core layout, is important and significant weight savings can be made by proper choice of core design and layout, bearing in mind the requirements of fatigue performance.

Compound costs are dominated by the two major constituents, i.e. the latex and the filler, and costs can in theory be minimised by manipulating the level of inorganic filler and the ratio of SBR to NR. This is not done without affecting the properties of the foam, however, and Table 10.1 shows the effect of different polymer ratios at 0 and 20 parts filler addition on some of the more important foam properties. The SBR latex used was a non-reinforced cold polymerised type, stabilised with fatty acid soap.

Latex foam products are always smaller than the mould in which they are made and the volume shrinkage figures in Table 10.1 show the magnitude of this effect compared with the original mould volume. The shrinkage is reduced as the proportion of SBR is increased. This means, of course, that it is not possible to take advantage of changes in the price structure of NR and SBR latices by simply changing the ratio in the compound. Too large a change results in an unacceptable change in the dimensions of the product and, to overcome this, new moulds would be required.

It can be seen that, with this type of SBR latex, product weights increase to maintain product hardness. It is possible to overcome this weight penalty by modulus reinforcement with polystyrene or high styrene resin latices added to the compound. A 10–20 per cent addition is normal. These resin latices, however, act as inert fillers and have the typical filler effect on physical properties. A better solution is to co-agglomerate the resin latex with the SBR latex at the latex production stage. This type of latex eliminates the weight penalty and may even show weight savings, without the deleterious effects on the physical properties of the foam.

Foam costs, therefore, are a combination of product weights and compound costs, which are not independent, but are dependent on the ratios of natural latex, SBR latex and filler and their relative prices. Minimising costs in a changing world, while maintaining acceptable foam and product properties, is a complex operation which can be, and has been, handled by a computer using multiple regression techniques.

TABLE 10.1
Effect of different polymer ratios (0 and 20 parts filler) on some foam properties

			NR:SBR:Filler			
	100:0:0	100:0:20	50:50:0	50:50:20	0:100:0	0:100:20
Weight index at constant hardness	100	99	105	103	110	107
Volume shrinkage (per cent)	16	15	15	14	14	12
Tensile strength (kPa)	120	68	56	42	38	36
Elongation-at-break (per cent)	290	180	250	140	130	130
Per cent hardness loss dynamic fatigue	27	40	30	40	21	24

Combustion behaviour of latex foam

The ignitability and the burning behaviour of plastics, particularly foamed plastics, has attracted a lot of attention in recent years. A variety of legal standards for foam cushioning in mattresses, automotive seating and furniture have been introduced in the USA and Europe and, although a detailed examination of latex foam in relation to these standards is not possible here, mention can be made of some of the more important features.

The combustion behaviour of latex foam should be considered under two headings, (1) flammability and (2) smouldering, because they are two different phenomena and proceed by different mechanisms.

FLAMMABILITY

Latex foam is a polymeric hydrocarbon in a low density form, intimately mixed with a large proportion of air and exposing a large surface area to it. As might be expected it ignites from an open flame and burns to completion.

The polymer is pyrolysed, to give low molecular weight volatile products which burn in the gas phase, generating more heat to pyrolyse more polymer. This results in a closed cycle of reactions which continues until the polymer is consumed. The use of alumina trihydrate as a filler reduces the burn rate of latex foam by endothermically losing water at the burning temperatures. Attempts to reduce the ignitability by, for example, large additions of halogenated fillers, such as PVC or PVDC together with synergistic agents such as antimony oxide, reduce the physical properties to unacceptably low levels.

Foam products are always covered in use, and can, of course, be protected from open flame ignition by suitable choice of a flame retardant cover.

SMOULDERING

Smouldering is a low temperature phenomenon such as is initiated by a lighted cigarette. Foam in the immediate vicinity of the cigarette is decomposed by the heat, by reactions which are exothermic, and a very delicately poised heat balance is set up. Depending on the conditions, the heat energy may be largely lost to the surroundings or may be available to decompose more polymer. In the latter case, further decomposition generates more heat energy and self-propagating charring occurs with the whole of the latex foam being consumed. Flaming ignition may or may not occur. The use of a sufficiently insulating cover on the foam will prevent the initial pyrolysis taking place and therefore gives protection. If the cover is inadequate to prevent this, it exacerbates the situation by partially insulating the foam and reducing the loss of the exotherm, leading to a more rapid promotion of self-propagating charring. The choice of the cover clearly needs to be made with care.

Filler type and quantity have very little influence on the phenomena of smouldering as also do antioxidants. The ratio of natural rubber, or more precisely cis-polyisoprene, to SBR, on the other hand, is critical. Latex foams containing more than 50 per cent polyisoprene do not exhibit self-propagating charring when tested in an uncovered state. For covered foams, the proportion of polyisoprene has to be increased to at least 75 per cent. The reason is largely due to the nature of the charred residue. SBR rubber leaves a rigid cellular char, which insulates the foam and prevents the exothermic heat escaping. Polyisoprene, on the other hand, melts and forms a sticky residue which retracts from the heat source.

Processing oils can help to solve the problem, but they have a softening effect, thus increasing product weights, and also have an adverse effect on other properties notably dynamic fatigue.

POLYCHLOROPRENE LATEX FOAM

While the bulk of the worlds latex foam is made from recipes based on natural rubber latex and/or SBR latex, there is a significant production of foam based on polychloroprene latex for applications where oil resistance is important or where improved resistance to ignition is required.

The processes used are those employed for standard latex foam production, but the recipes reflect the different characteristics of the latex. For example, the latex normally has a high pH, usually about 12·5, and contains fixed base as a stabiliser. It is usual to reduce the pH to 11·0 prior to gelation by the use of sensitising agents such as methyl diethanolamine. Both sodium and potassium silicofluoride can be used as gelling agents.

Curing systems are usually based on zinc oxide, with thiocarbanilide and ethanolamines (e.g. methyl diethanolamine) as accelerators.

The main differences in properties of polychloroprene foam are its very high volume shrinkage, its poorer hardness-to-weight ratio and its superior combustion characteristics which can be improved by the inclusion of fillers such as alumina trihydrate and antimony oxide.

Polychloroprene latex foam exhibits the phenomenon of self-propagating charring, similar to high SBR foams, probably because of the nature of the char residue.

In order to improve the hardness-to-weight ratio of polychloroprene foam and to enable curing to be accomplished at lower temperatures, DuPont introduced a process which uses MDI (diphenyl methane diisocyanate) as a crosslinker. The MDI is blended into the wet froth at the same time as the silicofluoride gelling agents. The product can be demoulded without distortion after 45 min at room temperature, but high temperature (120–140°C) curing and drying are necessary to produce the final properties.

In conclusion, this chapter has concentrated mainly on the technology of latex foam production and has deliberately avoided a detailed examination of the chemistry involved in the processes or the physics of the foam products produced. For this information, the reader is referred to refs. 1 and 2.

Typical formulations for the production of latex foams by the processes described are given in Appendix 1. Appendix 2 lists the ISO standards concerned with the testing of latex foams.

REFERENCES

1. MADGE, E. W., *Latex Foam Rubber*, Applied Science Publishers Ltd., London, 1962.
2. BLACKLEY, D. C., *High Polymer Latices – Their Science and Technology*, Vol. 2, Applied Science Publishers Ltd., London, 1966.

APPENDIX 1: FORMULATIONS

Typical formulations for the production of latex foam by the Dunlop (silicofluoride) process and the Talalay process are given below. A summary of the processing parameters is also given.

The natural latex is of the low-ammonia type and the SBR latex is of the polystyrene-reinforced, cold polymerised type. All figures given are parts by mass dry.

Dunlop process latex foam formulations

Natural rubber	100	50	0
Styrene–butadiene rubber } co-agglomerated	0	} 50	} 100
Polystyrene	0	} 8·75	} 17·5
Potassium oleate	1·5	1·2	1·0
Filler	20	20	20
Sulphur	2·5	2·5	2·5
Zinc oxide	6·0	6·0	6·0
Zinc diethyl dithiocarbamate	0·8	0·8	0·75
Zinc mercaptobenzthiazole	0·4	0·4	0·5
Diphenyl guanidine	0·6	0·6	0·5
Antioxidant	1·5	1·5	1·5
Sodium silicofluoride	2·0	2·3	2·5
Alkalinity	All mixes should be adjusted to contain 0·2 per cent ammonia		

Foam gel time	6 min
Mould cure	20 min
Drying/post-cure	75 min

Talalay foam

Styrene–butadiene rubber } co-agglomerated	} 100
Polystyrene	} 17·5
Potassium oleate	2·25
Process oil	3·0
Antioxidant	2·0
Sulphur	2·5
Non-ionic soap	0·1
Zinc oxide	5·0
Zinc diethyl dithiocarbamate	2·5
Mercaptobenzthiazole	1·2
Ammonia (35 per cent)	3·0
Potassium hydroxide	0·1

Vacuum expansion	2 min
Freeze	8 min
Carbon dioxide gassing	5 min
Low intermediate glycol	2 min
High intermediate glycol	2 min
Hot glycol (cure)	10 min
Drying/post-cure	75 min

Polychloroprene latex foam formulation (DuPont)

Polychloroprene	100
Dresinate	2·0
Non-ionic soaps	1·5
Zinc oxide	7·5

Antioxidant	2·0
Methyl diethanolamine	1·75
Thiocarbanilide	1·0
Alumina hydrate	20·0
Antimony oxide	4·0
MDI	15·0
Sodium silicofluoride	4·0
Gelling time	5 min
Mould cure (room temperature)	45 min
Dry/post-cure (120°C)	·180 min

APPENDIX 2: ISO STANDARDS RELATING TO LATEX FOAM

ISO 845 Cellular rubber and plastics — Determination of apparent density.

ISO 1798 Flexible cellular materials — Determination of tensile strength and elongation-at-break.

ISO 1856 Polymeric materials, cellular flexible — Determination of compression set.

ISO 1923 Cellular materials — Determination of linear dimensions.

ISO 2439 Polymeric materials, cellular flexible — Determination of hardness (indentation technique).

ISO 2440 Flexible cellular materials — Accelerated ageing tests.

ISO 3385 Flexible cellular materials — Test for dynamic fatigue by constant load pounding.

ISO 3386/1 Flexible cellular materials — Determination of compression stress/strain characteristics and compression stress value. Part 1 — Low density materials.

ISO 3582 Cellular plastic and cellular rubber materials — Laboratory assessment of horizontal burning characteristics of small specimens subjected to a small flame.

ISO 3795 Road vehicles — Determination of burning behaviour of interior materials of motor vehicles.

ISO 4651 Cellular rubber and plastics — Determination of dynamic cushioning performance.

Chapter 11

DIVERSE LATEX APPLICATIONS

T. D. PENDLE and A. D. T. GORTON
Tun Abdul Razak Laboratory, Brickendonbury, UK

NATURAL RUBBER LATEX THREAD

The term 'latex thread' normally refers to elastic thread produced by extruding compounded natural rubber latex through a fine nozzle into a bath of coagulant, then drying and vulcanising the product. The nature of this process demands a latex having excellent wet-gel strength, and NR latex is the only one that is suitable.

Latex threads can be produced in diameters ranging from 1·3 mm to 0·2 mm (known as 20 'count' and 125 'count', respectively), but the most common range is from 0·85 mm to 0·28 mm (30–90 counts).

The total production of latex thread is currently in the region of 30 000 tonnes per annum and is used mostly in the clothing industry. It is used for the production of elastic braiding and fabrics required for the 'stretch' parts of clothing. Prior to weaving into an elastic braid or fabric the latex thread is usually 'covered' (i.e. spirally wrapped) by a textile thread.

In addition to their uses in clothing, elastic fabrics made with latex thread are also used in footwear, e.g. in elastic-sided shoes, slippers and riding boots. Other applications of latex thread include the production of 'shock' cords for parachute harnesses and elastic cords for luggage strapping. In some countries open nets made from covered latex thread are even used, in place of string, for holding joints of meat together.

Compounding latex for thread production

Formulations for thread production are usually based on either centrifuged or creamed latex or blends of these. Creamed latex is preferred by some manufacturers, despite its higher cost, because of better processing characteristics but centrifuged latices are being increasingly used. The compounds described here are based on centrifuged latex but may be adapted for use with creamed latex.

A latex compound for use in the extrusion process should ideally possess a constant viscosity over several days storage together with the ability to gel rapidly at acid pH values. It should be free from air bubbles, coagulum, and poorly dispersed materials. In addition the compound should provide a vulcanised thread with a high modulus and low tension set and minimal staining in contact with copper. For satisfactory service life the thread should possess resistance to heat and washing. If the thread is to be used with white fabrics especially nylon, a good white colour is desirable. It is, in practice, not possible to obtain all of the above characteristics to an equal extent from one formulation. Most manufacturers, for example, produce both a standard grade and a 'heat-resistant' grade of thread and these are made from quite different compounds.

In respect of copper-staining, a compromise is normally necessary. Dithiocarbamates are the main cause of copper-staining, but it is not possible to omit these altogether without sacrificing rate-of-cure and, therefore, production speed. Consequently, compounds for thread usually contain a minimum quantity of dithiocarbamates, and have a thiazole (e.g. ZMBT) as the major accelerator. Also, it is usual to use zinc dibutyldithiocarbamate (ZDBC) rather than zinc diethyldithiocarbamate

TABLE 11.1
Latex thread formulations

	(a) Standard grade	(b) Heat-resistant grade
	parts by mass	
60 per cent NR latex, LATZ type	166.7	166.7
10 per cent potassium hydroxide solution	4.0	4.0
20 per cent potassium laurate solution	2.5	2.5
50 per cent sulphur dispersion	3.0	0.4
50 per cent ZMBT dispersion	3.0	2.0
50 per cent ZDBC dispersion	0.5	—
50 per cent TMTD dispersion	—	4.0
50 per cent ZDEC dispersion	—	4.0
50 per cent antioxidant dispersion	4.0	2.5
10 per cent thiourea solution	—	5.0
50 per cent titanium dioxide dispersion	10.0	10.0
50 per cent zinc oxide dispersion	3.0	4.0

Maturation: (a) 4 days at 25°C, (b) 3 days at 35°C.
Dry and cure: (a) 10 min at 140°C, (b) 15 min at 140°C.

(ZDEC) since the former is believed to be less prone to migration staining. Heat-resistant grades of thread are usually prepared using sulphur donors, such as tetramethyl thiuram disulphide (TMTD) which produces zinc dimethyldithiocarbamate as a by-product of vulcanisation. In such threads, therefore, copper-staining is not minimal. Formulations for standard and heat-resistant grades of thread are shown in Table 11.1.

In the preparation of latex compounds for thread production it is essential that all the dispersions are finely ground. It is recommended that the particles be less than 5 μm in order to avoid processing difficulties.

Equipment

Figure 11.1 is a schematic diagram of the basic features of an extruded thread plant.

The complete latex mix is stored in a 'header' tank controlled by a constant head device and moves from there to the manifold. From the manifold it passes through the glass capillary tubes into the acid bath. The gelled filaments are withdrawn from the acid bath by a set of stainless steel rollers and then pass into a washing bath and thence into the drying/vulcanisation oven. If the threads are to be formed into a

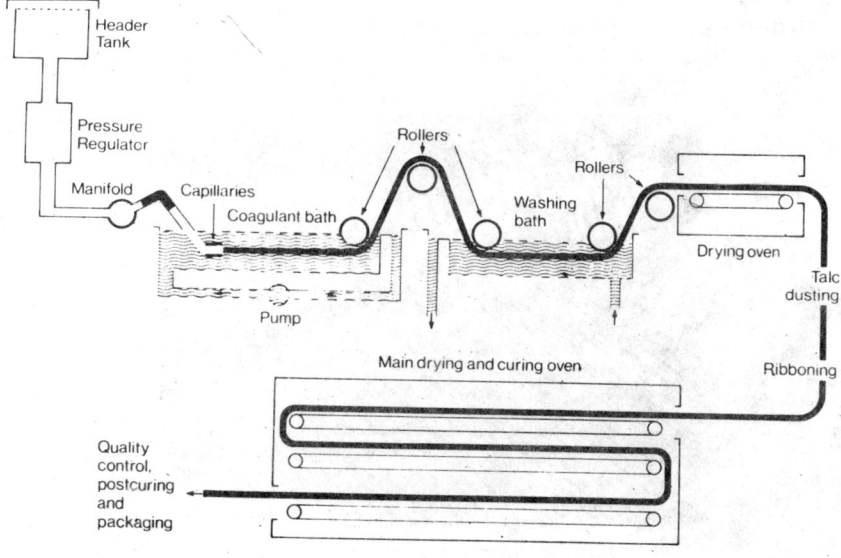

Fig. 11.1.

'ribbon' this is usually done part way through the vulcanisation stage. After emerging from the oven the threads may be subjected to extra heat treatment to complete vulcanisation and then inspected and packaged. Further details of the equipment are given below.

CONSTANT HEAD DEVICE

The latex compound is fed to the manifold via a constant head device in order to achieve a constant rate of latex flow. It should be noted here that a constant head only produces a constant flow rate if the viscosity of the latex compound remains unchanged with time and, in practice, mix viscosities do tend to change, so minor head adjustments may be necessary during a run.

A number of different designs are possible but, to be effective, any constant head device for this process should control the head to within ± 0.5 per cent or better and should be capable of applying heads ranging from 20–150 cm.

MANIFOLD AND CAPILLARIES

The manifold supplying the capillaries may be situated outside the acid bath or partially immersed in it. It is fitted with one or more air-vents to permit expulsion of air when filling with latex.

The capillary nozzles are attached directly to the manifold or connected to it by rubber tubing and are situated in the acid bath, approximately 2 cm below the surface, so that they are either horizontal

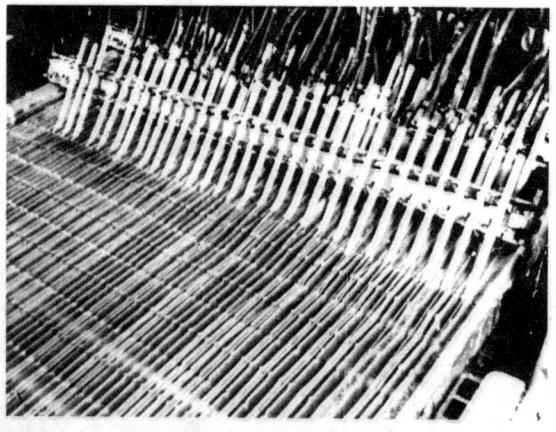

FIG. 11.2. View of extrusion capillaries and acid bath on a commercial thread plant.

or pointing slightly upwards. All of the capillaries attached to the manifold are matched both for length and bore otherwise unequal flow rates will occur. The ends of the capillaries from which the latex emerges are ground flat and free of chips or blemishes. Precision-bore glass tubing is recommended for the capillaries. Figure 11.2 is a photograph of the extrusion end of a commercial thread plant.

COAGULATING TANK

This vessel is made of material resistant to acetic acid and stainless steel is normally used. The length of the bath is sufficient to permit complete gelation of the thicker threads at the required speed of extrusion. The speed of extrusion, in turn, depends on the oven length available for drying and curing. Generally, acid bath lengths are 2–4 m.

The acetic acid in the bath is flowing in the reverse direction to the thread movement and circulating through a filter to remove any sediment which might otherwise build up on the threads. The acid strengths are determined at frequent intervals and further acid added as required.

The acid bath is fitted with separating devices to ensure that the gelled filaments cannot touch one another since any contact at this stage is impossible to correct without damage to the threads.

WASHING TANK

This is of stainless steel and of a similar length to the acid bath. The purpose of washing is to remove excess acetic acid and other hydrophilic materials from the threads. An excess of acid remaining in the thread would tend to slow the vulcanisation rate. It is usual to employ hot (60–80°C) water for washing as this gives more rapid extraction of ionic materials.[1] It is desirable that the water for the wash tank is free of copper and iron since these may cause discoloration of the thread.

DRYING/VULCANISING OVENS

The drying and vulcanisation stages occur concurrently in this process. The thread ovens are often divided into two or more temperature zones which increase from $c.$ 95°C at the beginning to 130–140°C near the end of the oven. The number of zones used is not particularly important and depends on the design of the oven; e.g. the number of zones may correspond to the number of belts used in a multi-pass oven. Multi-pass ovens are often used for thread production in order to achieve the oven length required with minimum use of floor space. A total oven length of about 100 m or more is necessary if high running speeds are to be attained. Thread ovens are normally heated via heat-exchangers using steam or gas. Air flow rates in the oven should not be so high as to cause movement of the filaments on the belts.

Ribboning
Ribbons are formed, usually from 40 separate filaments, by bringing the threads into longitudinal contact and applying a small compressive force by means of a roller. When properly prepared the individual threads in a ribbon adhere sufficiently well to withstand normal handling but may be easily separated without damage prior to covering. Fully vulcanised threads have insufficient residual tack to prepare a good ribbon and completely unvulcanised threads too much. Ribboning is therefore normally carried out part way through the drying and vulcanising process.

Latex compound processing
Thread compounds are matured prior to use by storing the compound at 20–30°C for a few days and an important part of this maturation is a degree of prevulcanisation. Maturation to swelling index values, in toluene, of between 9 and 6 g/g improves extrusion behaviour. Various methods of determining the degree of prevulcanisation are available and have been compared.[2] After maturation the latex mix is sieved prior to use — the presence of undispersed particles of curatives or of coagulum cannot be tolerated in this process. Many manufacturers pass the mix through a pressure homogeniser at this stage in order to ensure a completely uniform compound. If this is done it is important to sieve the compound after homogenisation. Following homogenisation and sieving the latex is stored in a closed vessel and the air pressure above the latex reduced, by a vacuum pump, to about 100 mm mercury in order to speed up the removal of air bubbles. Once all the air has been removed from the latex mix it can be fed, by gravity, to the constant head device that supplies the spinaret manifold.

Control of thread diameter
The main requirement of the process is to produce thread of a controlled constant diameter. The diameter of the thread made in the extrusion process depends on factors such as: the total solids content of the latex compound, the diameter of the capillary tube, the rate of flow of latex through the capillary tube (which is dependent on the diameter and length of the capillary), the viscosity of the latex and the pressure (from

the hydrostatic head) feeding the latex to the manifold, and the rate of pull-off of the thread by the rollers.

Pull-off rate is important since the extrudate is elastic in the wet-gel state and can be stretched quite significantly immediately after coagulation. The result of stretching is a reduction in thread diameter. This fact makes it possible to adjust the thread diameter by changing the speed of the draw-off roller. Once the conditions to produce a particular count are established it should only be necessary to make small adjustments to the head to compensate for any changes in mix viscosity with time. A mathematical relationship of the various factors controlling the diameter of the thread has been proposed.[3]

Testing of latex thread

The testing of latex thread is quite different from the testing of most other rubber products. The properties of most importance for thread are count (diameter), Schwartz value and elongation under a fixed load, although it is common to determine also the normal properties of tensile strength and elongation-at-break.

The diameter of the thread is determined indirectly, by measuring its density and then weighing a precise length (usually 1 m) of thread. The diameter is then calculated from these two results using the following equation:

$$D = 0 \cdot 02 \sqrt{\frac{m}{d}}$$

where m = mass of 1 m of thread (g), d = density of thread (g/cm^3) and D = average diameter of thread (mm).

The 'count' may then be readily calculated from the diameter using the following relationship:

$$\text{Count} = \frac{25 \cdot 4}{D}$$

The Schwartz value of the thread is the mean modulus, on extension and retraction during the sixth cycle, of a sample of thread which has been extended to and from an elongation 100 per cent greater than the elongation at which the modulus is required. For example, if a Schwartz value at 300 per cent elongation is desired, the thread sample is extended from zero to 400 per cent and back six times, and on the sixth cycle the load at 300 per cent is noted both on extension and retraction. The average of these two modulus

TABLE 11.2
Typical properties of latex threads[4]

	Standard type	Heat-resistant
Elongation-at-break (per cent)	525	525
Tensile strength (MPa)	30	30
Schwartz value 400/300 (MPa)	1·3	1·3
Accelerated ageing (per cent retained tensile strength after 14 days at 70°C)	>80	>80
Schwartz value (per cent retained after 4 h at 140°C)	—	>55

measurements is then the 300 per cent Schwartz value. Since latex threads cannot be held securely in the normal tensile machine grips it is usual to use special pneumatic grips or to use loops of thread made by knotting.

The elongation under load is, in effect, an alternative method of modulus measurement. The test measures the elongation produced when a predetermined load, usually 158 g/mm^2 or 280 g/mm^2, is applied to the thread. The tensile strength and elongation-at-break are determined on a tensile tester in the normal way, preferably using the special pneumatic grips.

In addition to the above measurements it is usual to determine the effect of oven ageing and also measure stress decay rates. An International Standard, ISO 2321 for the testing of latex thread has been published as well as various national standards including BS 903 parts H1 to H11, 1968; ASTM D2433-70; and DIN No. 7720. Typical thread properties[4] are given in Table 11.2.

RUBBERISED HAIR PRODUCTS

Rubberised hair is the name given to the high quality upholstery and packaging material made by bonding together a mass of curled animal hair with latex. Originally this product was made using only horse or hog hair but today most rubberised hair products are made from a mixture of animal hair and vegetable fibre, e.g. coir (coconut fibre). Rubberised coir products, made from vegetable fibre alone, are also in use and these are also described (inaccurately) as rubberised hair. In general the use of animal hair gives a more resilient product than does the use of vegetable fibres and automobile manufacturers often specify a minimum proportion of animal hair for material to be used in car upholstery. Automobile seating in fact represents the largest outlet for rubberised hair, which permits the design of very comfortable, lightweight seats which, because of the high permeability to air of the rubberised hair, provide cool seating even in hot weather. Rubberised hair and coir products are also used widely for the packaging of delicate instruments, because of their excellent shock absorbing characteristics and also for door-mats, gymnasium-mats and other similar applications.

Manufacturing techniques
The first step in the production of rubberised hair is the preparation of a sheet in which most of the fibres lie in the plane along which stresses will be applied during use. This can be done by hand but is normally carried out by machines designed for this purpose (Dr O. Angleitner, Linz, Austria; Dr E. Fehrer, Linz-Donau, Austria; O. Dilo, Eberback N, W. Germany). It is important at this stage that the fibres, whether animal or vegetable, have as much curl as possible in order to obtain the best properties in the final product.

After preparation of the fibre sheet it is then sprayed with the latex compound using traversing spray nozzles. Unless the pad is very thin it will not be possible to obtain adequate penetration of latex throughout the sheet by spraying from one side only, so it is usual to reverse the sheet after the first spraying and then spray the other side. The amount of latex applied varies to some extent with the intended application but it is usually sufficient to have a dry weight ratio of rubber to fibre of approximately 1:1. In general an increase in the rubber to fibre ratio produces an increase in load-bearing capacity and dimensional stability and a decrease in permanent set.

After completion of spraying the sheet is usually partially dried in an air oven at a relatively low temperature (60–70°C) to avoid excessive vulcanisation, and then cut to shape and vulcanised in a mould. Alternatively, the wet sheet may be shaped in perforated wire moulds and then dried and vulcanised. If rubberised-hair sheet is being produced, rather than moulded shapes, drying and vulcanisation may be carried out together in a hot air oven. In such a case, particularly where high air temperatures are used, it is necessary to ensure that the air-flow in the oven is directed through the thickness of the sheet rather than across its surface, otherwise over-vulcanisation of the surface may occur at the expense of the interior.[5]

Shaped rubberised hair products are sometimes produced by cutting sections from dried and vulcanised sheets and then sticking these together in the required manner. The latex compound used to spray the hair-pad is a satisfactory adhesive for this purpose

Formulations

Latex compounds for this application are fairly straightforward. They require good mechanical stability to resist the shear-stresses involved in spraying and they require a good degree of vulcanisation to ensure adequate product properties. It should be noted in respect of vulcanisation that animal hair tends to absorb sulphur whereas vegetable fibres do not, therefore formulations for animal hair based products must contain more sulphur than is otherwise necessary.

Formulations suitable for animal hair and coir fibre are given in Table 11.3. These are basic formulations to which fillers and pigments may be added as required. If fillers are used they should be of reasonably fine particle size and properly dispersed in the latex to avoid blockage of the spraying jets. It must be realised that the addition of more than about 25 phr of filler will result in an increase in compression set in the product. Some commercial products are made using blends of latices, e.g. NR/SBR or NR/CR blends but NR usually predominates.

For some applications rubberised hair products are required to have a measure of flame-retardance, e.g. for cars to be sold in the USA.[6] The achievement of good flame retardance is somewhat easier with animal hair products than with those based on vegetable fibres as the latter burn more

TABLE 11.3
Basic formulations for rubberised hair

	Animal hair	Coir or other vegetable fibre
	parts by mass	
60 per cent NR latex	166·7	166·7
25 per cent surfactant solution[a]	4·0	4·0
20 per cent potassium hydroxide solution	2·0	2·0
50 per cent antioxidant dispersion[b]	3·0	3·0
50 per cent ZDEC dispersion	2·0	2·0
50 per cent ZMBT dispersion	3·0	3·0
50 per cent sulphur dispersion	8·0	5·0
50 per cent zinc oxide dispersion	10·0	6·0
Water (distilled) — to 55 per cent total solids content		as necessary

[a] Sulphated ethylene oxide — alkylphenol condensates.
[b] For example, polymerised trimethyldihydroquinoline.
N.B. Dry at 60–70°C; cure 30 min at 100°C *or* dry and cure in high velocity air oven for 5 min at 140°C.

TABLE 11.4
Flame retardant formulations for rubberised hair

	100 per cent Animal hair	50/50 Hair/coir blend
	parts by mass	
60 per cent NR Latex	166·7	166·7
25 per cent surfactant solution[a]	4·0	4·0
20 per cent potassium hydroxide solution	2·0	2·0
50 per cent sulphur dispersion	7·0	6·0
50 per cent ZDEC dispersion	2·0	2·0
50 per cent ZMBT dispersion	3·0	3·0
50 per cent antioxidant dispersion[b]	3·0	3·0
50 per cent chlorinated wax dispersion[c]	10·0	30·0
50 per cent TBPA dispersion[d]	20·0	40·0
50 per cent hydrated alumina dispersion[e]	50·0	50·0
50 per cent antimony trioxide dispersion	4·0	12·0
50 per cent zinc oxide dispersion	6·0	6·0
Tetrasodium pyrophosphate	0·2	0·2

[a] Sulphated ethylene oxide — alkylphenol condensate type.
[b] For example, polymerised trimethyldihydroquinoline.
[c] Solid chlorinated paraffin wax (c. 70 per cent chlorine).
[d] Tetrabromo–bisphenol A.
[e] Hydrated alumina containing c. 30 per cent water.
Dry and cure 20 min at 120°C in hot air.
FMVSS burn rate (mm/min): 25–38 for 100 per cent animal hair and 50–75 for 50/50 hair/coir blend.

readily. However, recent work[7] has shown that coir can be treated to improve its flame retardance and this will clearly ease the compounding problems. The formulations and burning rates shown in Table 11.4[8] indicate what can be done by compounding the latex with flame retarding additives. Clearly, many other flame retardants and combinations of flame retardants can be used and further possibilities include the use of flame-retarding polymer latices (e.g. CR, PVC or polyvinylidene chloride) as blends with natural rubber latex.

CASTING PROCESSES

The term 'casting' refers to the process of forming hollow rubber products by depositing a layer of latex on the interior of a hollow mould. This may be done

Fig. 11.3. Toys made by casting with natural rubber latex. (Toys made by Garbep, Barcelona.)

in plaster-of-paris moulds where the plaster causes deposition of rubber, or in hot metal moulds using heat-sensitive latex mixes. These processes are used for the production of a range of children's toys, for film and stage-set parts, for advertising models and for meteorological balloons. Examples of toys made by this process are shown in Fig. 11.3.

Casting in plaster-of-paris moulds
This technique is very simple and can utilise a variety of latex compounds and thus permits the formation of products having a wide range of hardness and flexibility. It depends on the deposition of a layer of gelled latex on the mould surface due to absorption of water by the plaster and to destabilisation of the mix by calcium ions from the plaster. The procedure consists of filling the mould with the latex mix, allowing the filled mould to stand for a predetermined period to obtain the required deposition, pouring out the excess latex from the mould, and then drying and vulcanising the product in the mould.

Moulds for this process should be made from fairly fine plaster-of-paris powder,[9] using a ratio of approximately one part powder to two parts water. They should be made in two or more parts to facilitate removal of the product (and the master) and must necessarily have a filling hole — usually sited in the base of the product.

Latex mixes suitable for this process are detailed in Table 11.5.

TABLE 11.5
Mixes for casting in plaster moulds

Type of product	Soft	Semi-rigid parts by mass	Rigid
60 per cent NR latex	—	—	166·7
60 per cent prevulcanised NR latex	166·7	166·7	—
10 per cent surfactant solution[a]	—	—	5·0
50 per cent sulphur dispersion	—	—	4·0
50 per cent ZDEC dispersion	—	—	6·0
50 per cent zinc oxide dispersion	—	—	3·0
Filler[b]	—	75·0	300·0
Tetrasodium pyrophosphate	—	0·5	1·0
Water	—	27·0	90·0

[a] For example, casein solution.
[b] For example, china clay, whiting, etc.

Prevulcanised latices are clearly much simpler to use but are only suitable for unfilled or lightly filled mixes. Vulcanisable mixes, of course, may be used for any type of product if desired and are preferred for totally enclosed products. It is suggested that when preparing the compound the filler, water and tetrasodium pyrophosphate are mixed together and then added to the latex, thus avoiding the possibility of undispersed aggregates of filler particles in the latex. Prior to pouring into the mould the latex mix should be sieved and allowed to stand for a few hours to permit air bubbles to rise to the surface. Pigment dispersions may be added to the latex mixes if desired. Alternatively, the products can be painted, e.g. with plasticised cellulose lacquers for rigid articles or polyurethane lacquers for the more flexible ones.

Heat-sensitive casting

This process can be carried out using plaster moulds but it is much more efficient if metal, e.g. aluminium or stainless steel, moulds are used. The procedure can be similar to the 'slush-moulding' technique described above for plaster-of-paris but with the moulds heated to 70–80°C before filling, or it can be of the rotational casting type in which the closed mould, containing a carefully measured quantity of latex mix, is simultaneously rotated about two axes and heated to the required gelling temperature. The latter process is used in meteorological balloon production.

Similar latex mixes are used in both of these processes and suitable

TABLE 11.6
Heat-sensitive mixes for casting

	1	2	3	4
		parts by mass		
60 per cent NR Latex	—	166·7	—	166·7
60 per cent prevulcanised NR latex	166·7	—	166·7	—
25 per cent surfactant solution[a]	—	—	4·0	4·0
50 per cent sulphur dispersion	—	2·0	—	2·0
50 per cent ZDEC dispersion	—	2·0	—	2·0
50 per cent ZnO dispersion	2·0	2·0	1·0	1·0
25 per cent polypropylene glycol solution[b]	10·0	10·0	4·0	4·0
Distilled water	10·0	6·0	13·0	9·0

[a] Alkylphenol ethylene oxide condensate containing 9 moles ethylene oxide/mole.
[b] A molecular weight of about 700 g required.

examples are shown in Table 11.6. Mixes 1 and 2 have limited stability and should not be expected to be usable for more than 2–3 days when stored at 20°C. Mixes 3 and 4 have better stability and should be usable for at least 7 days if kept at 20°C.[10]

LEATHERBOARD MANUFACTURE

Leatherboard is the name given to the product made by binding ground leather fibres together with natural or synthetic latex, then compressing and drying the mass to form a sheet. Similar products are also made using cellulosic fibres. Natural rubber and CR latices predominate in this application although SBR latices are also used but mainly in blends with one or both of the other latices.

The predominant application for these materials is in the shoe industry where they are frequently used for forming the inner components of footwear.

Manufacturing techniques

Scrap leather, after careful removal of any metal, is first shredded using rotary cutters and then ground in a grinding mill. Cutting and grinding are usually done dry to avoid the discoloration that can occur, due to contamination with iron, if vegetable-tanned leather is ground wet. The properties of the finished board are largely determined by the length of the ground fibres: short fibres give harder more brittle boards, while long fibres increase flex resistance and tear strength.

The fibres are slurried in water at a concentration of about 5 per cent. The slurry is best prepared using a 'Hollander' but mechanical stirrers can also be used. A small amount of wetting agent may be added to the slurry to assist the wetting of slightly greasy fibres. Slurries of chrome-tanned fibre tend to remain fairly constant in pH (at about pH 6–7) but vegetable-tanned fibre slurries gradually reduce in pH to a value of 4 or less. As the pH value of the fibre slurry affects the stability of the latex, it is important that slurries of vegetable-tanned fibres be maintained at a constant pH throughout processing.

The stabilised latex is then added to the fibre slurry and thoroughly stirred in. The latex must be sufficiently stable to permit uniform mixing throughout the slurry without premature coagulation but must not be so stable as to make subsequent coagulation difficult. Its stability is controlled partly by the pH of the slurry and partly by the type and amount of added surfactant. It is usual to use non-ionic surfactants to ensure some stability in the latex at pH 6–7 and then effect coagulation by subsequent addition of aluminium sulphate. In some cases, if the balance of stability is just right, the latex will coagulate slowly after addition to the slurry without the use of a coagulating agent but usually the latter is necessary.

As indicated above the latex and slurry are preferably mixed at pH 6–7 but mixing may still be carried out even if the slurry pH has fallen to 4 or less provided: (1) that the mixing is sufficiently rapid; and (2) that the amount of non-ionic stabiliser is increased. At such low pH a coagulating agent is unlikely to be necessary. Alternatively, if rapid mixing cannot be carried out, the slurry pH should be raised by the addition of tetra-sodium pyrophosphate or sodium silicate.

After coagulation of the latex the slurry serum will have lost its turbidity and the slurry may then be fed to the filter moulds. In these moulds suction is applied to the mass via gauze or perforated bottom plates to remove much of the water. Wet sheets are removed from the filter moulds and plied prior to compression in a single-daylight press to squeeze out more water. The higher the applied pressure the harder and stronger is the final board but also the lower its permeability to water vapour. After pressing, the plied sheets are separated and finally dried at a temperature not greater than 55°C. They are then finished by calendering, moulding, embossing or lacquering as required. Specifications for leatherboard and other fibreboards are issued by SATRA[11] together with descriptions of test methods.

Formulations

These are necessarily very simple as all that is required is stabilisation and, optionally, protection against oxidation. A suitable formulation for natural latex is shown in Table 11.7.

TABLE 11.7
Leatherboard formulation

	parts by mass
60 per cent NR Latex	166·7
25 per cent surfactant solution[a]	1–2
50 per cent ZDEC dispersion	2

[a] Alkylphenol ethylene oxide condensates containing 15–30 moles ethylene oxide/mole.

LATEX CEMENT

Admixture of natural latex and cement was first reported in 1923[12] and since that time there has been a small but steady consumption of latex for this purpose. In recent years synthetic latices, e.g. PVA and XSBR types, have also been used. The addition of rubber to cement is believed to confer certain advantageous properties (e.g. flexibility, water resistance, frost-damage resistance, improved adhesion) to various substrates including steel, glass and old concrete. This has led to the use of latex/cement compositions for such applications as ship and bridge deckings, special-purpose floor screeds, repairs to old concrete and even for sticking the glass 'bubbles' in spirit-levels (glass to metal adhesion).

The degree of modification of cement or concrete properties depends on a number of factors, e.g. the ratio of rubber to cement, the proportion of rubber in the total mix — which normally contains sand and/or aggregates as well as cement, the degree of hydration achieved by the cement, and the proportion of air entrained in the product. In decking and floor screeds, where much of the water in the mix may be lost by evaporation and the hydration of the cement thereby inhibited, the products may be quite flexible. If the mix is applied in thicker layers, or if the cement is kept moist, hydration can proceed to completion and the product is more rigid.

Gazeley[13] has recently published a useful article describing the stabilisation of natural latex for use with cement and its effects on product

TABLE 11.8
Formulations for Portland cement (to give a polymer:cement ratio of 1:5)

	Natural latices		Synthetic latices	
	Evaporated concentrate	Centrifuged concentrate	PVA type	XSBR type
	parts by mass			
Latex	152	167	200	222
25 per cent surfactant solution[a]	20	20	—	—
10 per cent surfactant solution[b]	—	25	—	—
Portland cement	500	500	500	500
Sand	1500	1500	1500	1500
Water	150	125	100	90

[a] Alkylphenol ethylene oxide condensate containing approximately 30 moles of ethylene oxide/mole.
[b] Ammoniated casein solution or a polycarboxylic dispersing agent.

properties. He concludes that natural latex has virtually no effect on the ultimate degree of hydration of cement but does have a slight retarding effect on the rate of setting. According to Gazeley, one of the major effects of the addition of latex is to make the removal of air more difficult, and the proportion of air in the product has a major influence on strength properties.

Formulations suggested for various latices to be mixed with Portland cement are shown in Table 11.8. The same systems can be used with aluminous cements, although these have less destabilising effect on natural latex and therefore require less surfactants to be used. The natural latex formulations in Table 11.8 are prepared as follows: the latex and surfactants are mixed together, then the sand/cement/water, and finally the two parts are combined. A similar technique can be used with the synthetic latices although in these cases the whole formulation may be mixed in one stage if desired. If high strength properties are required in the product it is suggested that an antifoaming agent (e.g. trimethyl hexanol) is added to the cement paste prior to the addition of the stabilised latex.

It can be seen from Table 11.8 that the synthetic latices have the advantage of simplicity of compounding and they also show superior stability when mixed with cement. Natural latices have the advantage of conferring superior flexibility on the product.

MISCELLANEOUS APPLICATIONS OF NR LATEX

Over the years a large number of miscellaneous applications involving NR latex have been proposed and some are currently in operation. In this section these processes will be briefly described.

Extruded tubing

High quality rubber tubing can be made using a simple apparatus (Fig. 11.4) and a heat-sensitive NR latex compound. The latex compound is stored in a tank and kept at a temperature of 20°C or less. It flows via a constant level device (A) to the extruder. This normally is made of concentric polished glass tubes fitted with a cold water jacket (D) at the top and a hot water jacket (E) below. When the latex compound enters the heated zone around the hot jacket, it gels in the annulus between tubes B and C and is slowly extruded from the bottom of the apparatus. The cold jacket is maintained at 15–20°C to stop the compound gelling

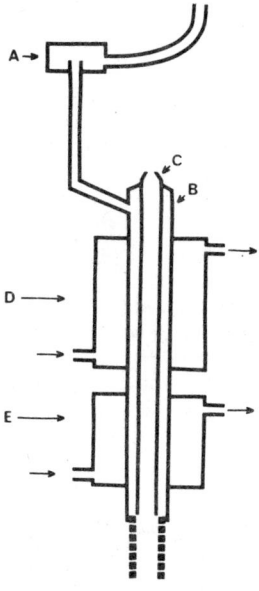

Fig. 11.4.

in the supply tube. The hot jacket is normally at 50–70°C depending on the cross-section of the tubing that is required.

The tubing in the form of a wet-gel is passed through a detackifying bath of a talc slurry or other material, followed by a wash bath for leaching, and is dried and vulcanised in a hot air oven.

The rate of extrusion is controlled partly by the hydrostatic head of the latex compound reservoir above the extruder, but also to some extent by the length of gelled tubing hanging from the extruder. Typically, a 50–60 cm head with a 50 cm hanging length will give extrusion rates of 20–30 cm/min. Although the extrusion rate is slow the simplicity of the apparatus allows a number of extruders to be run concurrently from the same reservoir of latex compound. A compound formulation is given in Table 11.9. It is possible to prepare tubing from a prevulcanised latex using a similar formulation, omitting the curatives, and ensuring the pH is at 7·5–8·0 by adjustment of the level of formaldehyde.

Latex/siliconate mixtures

A combination of latex concentrate with monomethyl sodium siliconate functions as a water repellent[14] and has found widespread application in

TABLE 11.9
Heat-sensitive extrusion formulation

	Parts by mass
60 per cent NR latex, LATZ type	166·7
25 per cent non-ionic stabiliser solution[a]	1·0
40 per cent formaldehyde solution	2·2
Water	30·0
50 per cent sulphur dispersion	2·5
50 per cent ZDEC dispersion	2·0
50 per cent zinc oxide dispersion	2·0
50 per cent antioxidant dispersion	1·0
10 per cent polyvinylmethylether solution	20·0

[a] Alkylphenol ethylene oxide condensate containing 30 moles ethylene oxide/mole.

buildings which suffer from dampness, especially older structures built without dampcourses and subject to rising damp. The latex siliconate mixture is pressure injected into the masonry through a series of holes, thus forming a permanent dampcourse without expensive structural alteration. It is likely that the siliconate solution itself penetrates the finer pores in the wall, whilst the larger pores are progressively blocked by the aggregation of rubber particles. The presence of calcium ions in the brick or stone will assist in the deposition of the rubber from latex. The mixture of latex and siliconate can also be used to waterproof floors and walls by surface application. Suitable formulations are given in Table 11.10.

TABLE 11.10
Damp-proofing formulations

Additive	Dampcourse	Surface waterproofing
	parts by mass	
40 per cent aqueous siliconate[a]	8·0	8·0
60 per cent NR latex	80·0	12·0
Water	12·0	80·0
Alumina cement	5·0	5·0

[a] For example, Dow Corning 722.

Latex rubber sheeting

Techniques have been devised and are used commercially for the continuous production of thin films or sheets of vulcanised latex rubber.[15,16] In one method,[15] an NR heat-sensitive compound is deposited on an endless steel or fabric conveyor belt, which carries the compound through heating zones where it is gelled, dried and vulcanised. By this means, latex films of up to 4 m width are manufactured in continuous lengths. Latex sheeting is used for a wide variety of applications, including dental uses, waterproof clothing, etc.

Latex bitumen mixtures

For many years rubberised bitumen has been used as a road surfacing material.[17] The presence of the rubber, added as latex, produces marked changes in the strength and flow properties of the bitumen. Its low temperature brittleness and high temperature deformability are reduced, whilst its elasticity, toughness, tack and adhesion are increased. These properties enable asphalt road surfacings to be obtained with improved wear characteristics and greater resistance to cracking.[18,19] The proportion of rubber to bitumen usually varies from 1–5 per cent. Various latex types can be used, including evaporated NR latex.

Rubberised bitumen may be prepared by adding latex to straight-run bitumen heated to 150°C. Under these conditions the water flashes off and the rubber blends in with the bitumen. Small amounts of a silicone anti-foaming agent may be needed to prevent excessive foaming. Alternatively, latex can be mixed with cut-back bitumen by blending at 70°C, and raising the temperature to 100°C to remove the water. The mixture is then ready for use as a masterbatch for addition to bitumen. Whichever method is used, it is essential that the temperature does not rise beyond 150°C for any prolonged period, otherwise the rubber will be oxidised and its effectiveness reduced.

Extruded tapes

One method of making tapes is to recombine a number (e.g. 4–6) of individual extruded threads. This differs from the 'ribboning' process described in the first section of this chapter in that the threads are brought into contact prior to vulcanisation to ensure a permanent bond. It has been shown at MRPRA, however, that tape can be extruded directly by using a glass or polished steel slit die on an extruded thread plant (Fig. 11.5).

Rubber tapes are preferred to woven threads for certain applications in the clothing industry.

FIG. 11.5. Tape extrusion dies compared with a thread extrusion capillary tube.

Sports surfaces

Polymer-based sports surfaces are claimed to have several advantages over the traditional ones of grass, cinders or concrete, including lower maintenance costs, improved resistance to mechanical damage and weathering, and improved bounce and safety. Some commercially available materials consist of tyre crumb with natural rubber latex as binder,[20] and have the merit of being relatively inexpensive and easily poured into place on site to provide a seamless rubberised surface. The formulation given in Table 11.11 is for a variation of this type, in which Portland cement is added to the mix to aid drying, thus giving good setting characteristics and reducing drying time — an important feature when surfaces have to be laid in cold or humid conditions. The mix is prepared as a two-part, self-setting composition, the tyre crumb and cement being mixed into the stabilised latex on site.

TYRE-CORD DIPS

The reinforcing cords used in the construction of pneumatic tyres are frequently precoated with a rubber composition designed to ensure good adhesion between the cord and the tyre carcass. The rubber mix is applied to the cord by immersing it in a latex compound and the coated cord is then dried prior to use in tyre building. The rubber applied to the cord does not usually contain vulcanising agents but is believed to be vulcanised in the tyre due to diffusion of curing agents from the carcass compound. Originally natural latex based compounds were used for this purpose but in recent years this has largely been replaced by synthetic

TABLE 11·11
Sports-surface formulation

	parts by mass
60 per cent NR latex	166·7
25 per cent surfactant solution[a]	8
Vulcanising/protectant paste (composition as given below)	68
Tyre crumb	175
Portland cement	150
Vulcanising/protectant paste:	
50 per cent sulphur dispersion	3
50 per cent ZDBC dispersion	3
50 per cent ZMBT dispersion	3
50 per cent zinc oxide dispersion	2
50 per cent antioxidant dispersion[b]	4
50 per cent UV absorber dispersion[c]	4
Thiourea	1
25 per cent surfactant solution[a]	8
20 per cent glycine solution	15
10 per cent thickener solution[d]	5
10 per cent casein solution	20

[a] Alkylphenol ethylene oxide condensate containing 30 moles ethylene oxide/mole.
[b] For example, polymerised trimethyldihydroquinoline.
[c] For example, benzotriazole derivatives.
[d] Polyacrylate type.

latices which exhibit better adhesion to modern cord materials, e.g. nylon, polyester, steel, etc.

Most tyre-cord dipping mixes today are based on three components: a resorcinol–formaldehyde resin, a vinyl pyridine–styrene–butadiene terpolymer latex, and a styrene–butadiene latex. In some cases, however, the first two components alone are sufficient. Typical formulations for tyre-cord dipping mixes are shown in Table 11.12.

In the coating of tyre-cords the cords are drawn around a roller immersed in the latex mix, then through squeeze rollers to remove excess latex, and then into a drying oven where the water is evaporated and the condensation of the resin completed. A final temperature of 130°C is necessary to complete the resin condensation and ensure adequate bond strength.

The adhesion obtained with dipped tyre-cords depends on a number of factors, e.g. the resin–rubber ratio, conditions of vulcanisation, type of

TABLE 11.12
Tyre-cord dipping formulations

	Nylon cord	Polyester cord
	parts by mass	
Vinyl-pyridine terpolymer latex (41 per cent TSC)	285	428
Styrene–butadiene latex (40 per cent TSC)	145	—
Resorcinol–formaldehyde resin solution	465[a]	500[b]
Distilled water	105	72
	Mature 6 h at 25°C	Mature 6 h at 25°C

[a] Prepared as follows: Resorcinol, 11·0; 40 per cent formaldehyde, 15·0; Sodium hydroxide, 0·3; Water, 240; Mature c. 4 h at 25°C.

[b] Prepared as follows: Resorcinol, 11·0; 40 per cent formaldehyde, 7·5; Sodium hydroxide, 0·3; Water, 198·0; Mature 6 h at 25°C.

latex used, nature of the tyre cord, amount of dipping mix picked up on the cord, and the distribution of the adhesive within the cord. In addition to these factors, the measured bond strength will depend greatly on the conditions of test, e.g. on test temperature and humidity.

REFERENCES

1. PENDLE, T. D., *NR Technol.*, 1974, **5**(2), 21.
2. GORTON, A. D. T. and PENDLE, T. D., *NR Technol.*, 1976, **7**(4), 77.
3. JAMES, R. G., *Trans. IRI*, February 1949, **24**, 220.
4. COLLINS, J. L., Malaysian Rubber Producers' Research Association, Brickendonbury, Hertford, England, unpublished work.
5. POLE, E. G. *Rubb. Dev.*, 1959, 12, 11.
6. FMVSS 302, *Flammability of Interior Materials*, US Department of Transportation.
7. GREEN, J. H. S. and JARMAN, C. G. (eds.), *Proc Workshop on the Flame Retardant Treatment of Rubberised Coir*, The Tropical Products Institute London, 1980.
8. MRPRA, *Technical Information Sheet L22*, The Malaysian Rubber Producers' Research Association, Brickendonbury, Hertford, England, 1978.
9. MRPRA, *Technical Information Sheet L36*, Malaysian Rubber Producers' Research Association, Brickendonbury, Hertford, England, 1979.
10. GORTON, A. D. T. and PENDLE, T. D., *Proc. Intern. Rubb. Conf.*, Venice, 1979, 161–74.

11. Shoe and Allied Trades Research Assocation, Satra House, Kettering, Northants, England, 1959.
12. CRESSON, L., BP 191 474, 1923.
13. GAZELEY, K. F., *NR Technol.*, 1979, **10**(2), 29.
14. HURST, H., BP 848 352, 1960.
15. Four D Engineering, BP 1 326 541 1973.
16. Dunlop. BP 957 014, 1964.
17. THOMPSON, P. D., Natural rubber in road surfaces, *NR Tech. Bulletin*, No. 9, Malaysian Rubber Producers' Research Association.
18. MULLINS, L., *NR Technol.*, 1971, **2**(3), No. 7.
19. TEBBUTT, D., *The Queens Highway*, April 1973, **100**, 4.
20. ANON., *NR Technol.*, 1975, **6**(2), 39.

INDEX

ABS plastics, 35
Acrylic adhesives, 123, 134
Acrylic latices, 33–5
   copolymerisation, 34
   glass transition temperatures, 33
   monomer reactivity ratios with
      styrene, 33
   polymerisation, 34
   solids contents, 35
Acrylonitrile–butadiene, 27
Acrylonitrile determination, 61, 62
Additives, 171
Adhesives, 28, 38, 119–43
   bottle labelling, 124
   building industry, 126–35
   casein modified, 124
   ceiling tile, 133
   cold seal, 120–1
   crosslinking PVA, 140
   dispersion-based, 121, 124
   envelopes, 124–5
   flooring, 128–32
   footwear, 126
   for boxes, cartons and cases, 121–4
   handling considerations, 119–20
   selection of, 123
   wall cladding, 132–3
   wood, 135–42
      veneering, 142
Aeration, 5
Agglomeration, 37
Agitation, 5

Alkalinity, 53
Alkalis, 6
Alkali-soluble polymers, 159
Alumina trihydrate (ATH), 75
Ammonium acetate system, 82
Antioxidants, 7
Antiozonants, 7
Aqueous pigment coating, 93
Artificial latices, 44–5
Asbestos fibres, 105

Balloon manufacture, 201–2
Barrier coatings, 40–1
Bed tickings, polymer treatment, 116
Belting, 38
Binders, 93–117
   impregnation process, 110–11
   latex types, 98–9
   paints, 146
   paper coating, 97–8
   print bonding, 109–10
   sole latex, 98
   spray bonding, 113
Biocides, 162
Bitumen, 249
Bitumen/rubber emulsions, 128
Bond strength development rate, 137
Bonded fibre fabrics, 109–13
Bonding agents, building industry,
      134
Boric acid content, 66

Bottle
  labelling, adhesives, 124
  teats, 202
Building industry
  bonding agents, 134
  sealants, 134–5
  see also Adhesives; Binders; Paints
Butadiene–acrylonitrile, 29
Butyl latex, 44
Butyl rubber, 44

Carbon dioxide, 5, 55
Carbon dioxide number, 14, 57
Carboxylic acids, 100
Carboxylic–acrylic copolymers, 98
Carboxylic–acrylic latices, 98
Carboxylic–styrene–butadiene copolymers, 101
Carboxylic–styrene–butadiene latex, 29–31, 71, 87, 93, 98, 111
Carpet applications, 71–91
  backing systems, 72
  foam
    application technique, 80–2
    backing, 80–91
    properties, 90–1
    quality aspects, 90
    system selection, 90–1
  high solids froth secondary backing, 78–80
  jute backing, 76
  market statistics, 71
  non-gel foam process, 87–9
  precoating
    compounds, 73
    formulations, 75
    function, 75
    process, 72–6
  primary backings, 76
  secondary backings, 76–80
  sodium silicofluoride gelation process, 85–7
  unitary backings, 76
  zinc ammine gelation system, 82–5
Casting processes, 238–42
  heat-sensitive, 241
  plaster-of-paris moulds, in, 240–1

Cationic deposition aids, 104
Ceiling tile adhesives, 133
Cellulose
  ether, 158
  fibres, 104–5
  film, 121
Cement, 244–6
Centrifugation, 11
Chain transfer agent, 24
Chemical stability, 58–60
  natural rubber latex, 59
  synthetic rubber latices, 60
Chlorination, 199
Coagulum
  content, 51
  test, 183
Coalescents, 161–2
Codification of synthetic rubber latices, 49–50
Colloids, 158–60, 164, 171
Combustion characteristics, 225
Compounding ingredients, 6–9
Compounding process, 9–10
Concentration processes, 11
  test methods, 51–3
Construction industry, sealants, 127–8
Contamination prevention, 4
Copper, 14, 66
Coulter Counter, 65
Creaming, 5, 11, 17, 47, 65, 229
Crosslinking PVA wood adhesives, 140–2

Damp-proofing formulations, 248
Defoamers, 162
Dehydration, 5
Density, 14–15, 53
Dextrine, 124
Dibutyl phthalate, 136
Diphenyl guanidine (DPG), 87
Dipping processes, 173–205
  applications, 173–4
  balloon manufacture, 201–2
  basic principle, 174
  beading, 188
  bottle teats, 202

INDEX

Dipping processes—*contd.*
 compound requirements, 187
 compounding, 176–81
 compounding aims, 176
 cure testing, 203
 dry stripping, 188–9
 drying, 187–8, 190
 former
  cleaning, 185
  design, 175–6
 fundamentals, 174–6
 future developments, 204
 glove manufacture, 191–200
 leaching, 188
 light, 204
 natural rubber latex, 176–7
 nitrile latices, 179–80
 ozone resistance, 204
 polychloroprene latices, 180–1
 prevulcanised compounds, 177
 progress in, 173
 protectives, 184–91
 solid/liquid interface, 174
 solvent and chemical resistance assessment, 204
 solvent stripping, 190
 soothers, 202
 stripping methods, 188–90
 synthetic rubber latices, 177–81
 tank construction, 185–6
 technique, 187–8, 190
 testing methods, 183, 203–4
 tyre-cords, 250–2
 unprevulcanised compounds, 177
 vertical dipping, 175
 vulcanisation, 188, 190
 wet stripping, 189
Dispersing agents, 7, 8, 160–1, 171, 210
Dispersion formulation, 8
Dithiocarbamates, 230
'Dry picking', 96
Dry polymer content, 52
Dry rubber content, 52
Dunlop process, 227

Electrodecantation, 11
Electron-microscopy, 65

'Elephant hide', 86
Emulsifier formulation, 9
Emulsion polymerisation, 21–4
 advantages, 22
 basic ingredients, 23
 consumption statistics, 22
 description of process, 23
 disadvantages, 23
 initiator, 24
 monomers, 24
 soap (emulsifier), 24
Emulsion polymers, 154
Emulsions, 145
End-point detection, 55
Envelope adhesives, 124–5
Ethylene copolymers, 42
Evaporation, 11, 17–18
Extenders, 171
Extruded tapes, 249
Extruded tubes, 246–7
Eye protectors, 6

Fillers, 7, 9, 73, 75, 79, 83, 84, 138, 222, 225
Flocculation, 55
Flooring, adhesives, 128–32
Foam, 207–28
 backing. *See* Carpet applications
 batch operation, 212
 combustion
  behaviour, 224–5
  characteristics, 225
 compression set, 220
 continuous operation, 212–13
 costs, 221–2
 curing, 214, 218
 demoulding, 212, 218–19
 drying, 215, 219
 Dunlop process, 208–15
 elongation-at-break, 220
 fatigue properties, 221
 flammability, 224
 formulations, 226–8
 gelling, 218
 gelling agents, 211–12
 gelling reactions, 213
 hardness, 220

Foam—*contd.*
  hardness-to-weight ratio, 225–6
  history, 207
  mould temperatures, 213–14
  moulding, 212–13, 216–18
  outlets, 207
  polychloroprene latex, 225–7
  polymer chemistry, 208
  preparation, 209–10, 216
  properties, 220–6
  smouldering, 224
  standards, 228
  surface chemistry, 208
  Talalay process, 208–9
  tensile strength, 220
  volume shrinkage, 222, 225
  washing, 214–15, 219
Foamers, 79
Foaming process, 211, 216, 218
Footwear adhesives, 126
Formulation calculations, 10
Fractional creaming technique, 65
Freeze–thaw stability, 18
Freeze–thaw stability test, 66
Frothing, 74
Furniture adhesives, 135–42

Gel
  range, 85
  sensitisers, 7–8
  stabiliser, 87
Gelation technique, 82, 86–7
Gelling agent, 85, 211–12
Gelling reactions, 213
Glove manufacture, 191–200
  alcoholic coagulant, 194
  aqueous coagulant, 193
  beading, 195
  chlorination, 199
  coagulant bath, 193–4, 196
  drying, 195, 196, 198, 199
  electricians, 199
  former
    cleaning, 192
    design, 192
    pre-heat tank or oven, 193

Glove manufacture—*contd.*
  houseware, 196–9
  leaching, 195, 198, 199
  second dip, 198
  stripping, 196, 198
  supported, 200
  surgeons, 191–6
  tank requirements, 194–5
  vulcanising, 196, 198, 199
Gravure printing, 96, 97
Gross particle content, 51

Handling, 4
Heat
  sensitiser, 8
  sensitivity evaluation, 183
Heveaplus MG latex, 18
High-ammonia latex, 12, 13, 17
Higher fatty acids, 58
Hydrofluoric acid, 85

Impregnation process, binders, 110–11
Initiator systems, 24, 40
Iron, 14, 66

LABA latex, 12, 66
Labelling, bottle adhesives, 124
LAPCP latex, 12, 14
LATD latex, 12
Latex
  applications, 3
  cement, 244–6
  consumption statistics, 3
  definition, 1
  foam. *See* Foam
  properties, 2–3
  thread, 229–36
    coagulation tank, 233
    compounding, 229–31, 234
    constant head device, 232
    copper-staining, 230
    diameter control, 234, 235
    drying vulcanising ovens, 233
    elongation under load, 236

Latex—*contd.*
  thread—*contd.*
    equipment, 231–4
    formulations, 231
    heat-resistant grades, 230–1
    manifold and capillaries, 232–3
    modulus measurements, 235–6
    properties, 236
    ribboning, 234
    testing methods, 235
    washing tank, 233
    types, 1–2
Latex/siliconate mixtures, 247–8
LATZ latex, 13
LAZDC latex, 13
LAZN latex, 13
Leatherboard, 242–4
  application, 242
  formulations, 244
  manufacturing techniques, 242
Letterpress process, 95–7
Lick roll, 74
Lithography, 95–7
Low-ammonia latex, 12–16

Magnesium, 14
Manganese content, 66
Mechanical stability
  natural rubber latex, 54–5
  synthetic rubber latices, 56
Melamine formaldehyde resin, 87
Metallic elements, 14
Methyl methacrylate, 18

Natural rubber latex, 2, 3, 5, 9, 11–20, 229–36
  applications, 19–20
  centrifuged, 12, 19, 47, 48
  chemical stability, 59
  creamed, 17, 47
  dipping process, 176–7
  evaporated, 17–18, 48
  mechanical stability, 54–5
  miscellaneous applications, 246–50
  non-rubber solids, 15
  prevulcanised, 19

Natural rubber latex—*contd.*
  quality control, 56–8
  solids content, 48
  speciality, 18–19
  specifications, 47–9
  twice-centrifuged, 16
Nitrile latices, 38–9, 179–80
  dipping characteristics, 181
  modified type, 39
  polymerisation, 38
  reactivity ratios, 39
  uses, 39
Nitrogen content, 66
Non-gel foam process, 87–9
Non-rubber solids, 15, 16, 18
Non-volatile acid, 15, 16, 58
Non-woven fabrics, 106
Nonylphenol-ethylene oxide, 160

Packaging, adhesives, 120–5
Paints, 145–67
  additives, 171
  adhesion, 163–5
  binders, 146
  biocides, 162
  chalking, 158
  chemical resistance, 153
  coalescents, 161–2
  colloids, 158–60, 164, 171
  constitution, 145–7
  critical pigment volume concentration (CPVC), 147, 155–7
  deficiencies in, 163–4
  definition, 145
  defoamers, 162
  design, 149–56
  developments, 164–6
  dispersants, 160–1, 171
  extenders, 156–8, 171
  film, 152
  flame-retardant, 165
  flow properties, 159, 163, 164
  formulations, 168–71
  gloss, 164, 168
  ingredients, 146
  manufacturing procedure, 168

Paints—*contd.*
  matt, 165, 170
  minimum film-forming temperature (MFT), 150–2, 161, 162
  miscellaneous additives, 162–3, 166
  monomer composition, 149–53
  opacity, 155
  particle size, 154–6
  pigment(s), 156–8
    binding properties, 156
    volume concentration, 147
  pigment/resin technology, 166
  plastic pigments, 165
  polymer film hardness, 149–50
  properties, 145–6, 149–56
  PVC, 164, 165
  rheology, 163
  solvent-based alkyd, 163
  stability, 159
  storage stability, 154
  thermoplasticity, 164, 166
  thixotropic, 159, 163
  titanium dioxides, 171
  types of, 147–9
  vinyl silk latex, 169
  viscosity stability, 159
  water
    phase, 153–4
    resistance, 154
Paper coating, 93–101
  binder(s), 97–8
    function, 94
    monomer ratio, 99–100
      polymer composition, 99
      variables influencing coating performance, 101
    products available, 101
  description of process, 94
  formulation of mix, 97
  printing process, in, 95–6
  requirements, 96
Paper making process, 102–9
  fibres and applications, 104–6
  latex
    addition method, 102
    types used, 102
  saturation latices and applications, 108

Paper making process—*contd.*
  saturation process, 107–8
  wet-end addition, 103, 108
Papermaker's alum, 103–4
Particle
  properties, test methods, 62–5
  size, 35, 37, 65, 154–6, 210
pH
  determination, 183
  effects, 8, 31, 37, 64, 136, 140, 153, 160, 181, 225, 230, 243
  measurement, 53
Plaster-of-paris moulds, 240–1
Plasticisers, 136, 161
Polychloroprene latices, 42–4, 180–1
  dipping characteristics, 181
  foam, 225–7
  general purpose, 43
  properties, 43
  reactivity ratios, 43
  solids contents, 43
  speciality, 43
  uses, 43–4
cis-1-Polyisoprene, 44
Polymer composition test methods, 60–1
Polyphosphates, 73
Polysiloxane emulsion, 84
Polystyrene latices, 35
Polysulphide sealants, 134
Polyurethane
  foam, 207
  latices, 44, 45
Polyvinyl acetate latices, 31, 98, 123, 124, 134–7, 140–2
Polyvinyl alcohol, 98, 138
Polyvinyl chloride latices, 41–2
  applications, 42
  preparation, 42
  reactivity ratios, 41
  solids contents, 42
Polyvinylidene chloride latices, 39–41, 98, 120
  initiator systems, 40
  polymerisation, 40
  properties, 40
  uses, 40–1
Portland cement, 246

INDEX

Potassium, 14, 17
  hydroxide, 18, 131
  number, 14, 17, 48, 49, 56–7
  polyphosphates, 160
Prevulcanised natural latex, 19
Print bonding, binders, 109–10
Printing inks, 96
Printing processes, 95–6
Propafilm C, 121
Propylene glycol, 161
Protective colloids, 6
Protectives
  general observations, 191
  high output manufacturing plant, 190
  low output manufacturing plant, 184–90
  manufacturing cycles, 184–91
Pumps and pumping, 4, 119

Quality control, natural rubber latex, 56–8

Redox systems, 36
Resin emulsions, 130–1
Rubber latex adhesives, 129
Rubber/resin emulsions, 128–9
Rubberised hair products, 236–42
  flame-retardance, 238–9
  formulations, 238–9
  manufacturing techniques, 237

Safety precautions, 6
Sampling procedures, 51
Saturation process, 107–8
Scraper blade, 74
SDEC/ZnO latex, 12
Sealants
  building industry, 134–5
  construction industry, 127–8
  water-based, 134
'Seeding' technique, 31
Shear, 4
Sheeting, 249
Silicone sealants, 134

Skinning, 5
Sludge content, 58
Soap, 24, 35, 38
  adsorption methods, 65
  content determination, 64
  deficiency determination, 64
Sodium
  lauryl sulphate, 129
  polyphosphates, 160
  silicofluoride (SSF), 37, 38, 208, 212
  gelation process, 86–7
Solution emulsification technique, 44
Solution polymerisation process, 44
Soothers, 202
Specific heat capacity, 14
Specifications
  natural rubber latex, 47–9
  synthetic rubber latex, 50
Sports surfaces, 250
Spray bonding, binders, 113
Spraying, 6
Stabilisers, 6, 18
Stability, 1, 4, 18
  chemical, 58–60
  mechanical, 13, 48, 54–6
Standards, 68–9, 228
Stirring, 5, 10, 55, 56
Storage, 4, 10
Styrene–acrylic, 29
Styrene–butadiene latex, 2, 9, 27–9, 35–8, 72, 84, 86, 88, 89, 99–101, 129, 132, 134, 209, 210, 222
Styrene–butadiene/natural rubber, 210
Styrene–butadiene rubber latices, volatile unsaturates, 61–2
Styrene content determination, 60
Surface-active agents, 6
Surface coatings, 28
Surface free energy, 63
Surface tension determination, 63
Synthetic fibres, 106
Synthetic latex binders. *See* Binders
Synthetic rubber latices, 21–46
  batch reactions, 25–6
  chemical stability, 60
  codification of, 49–50

Synthetic rubber latices—*contd.*
 complete conversion, 28
 continuous reaction, 26–7
 continuous stirred tank reactor (CSTR) system, 27
 dipping process, 177–81
 future developments, 45–6
 mechanical stability, 56
 polymer composition determination, 60
 polymerisation procedure, 31
 post-reaction treatment, 28
 preparation, 25–9
 properties, 25
 semi-continuous reaction, 26
 specifications, 50
 tube system, 27
 types, 29–45
 uses, 28–9

Talalay process, 215–19, 227
Tape extrusion, 249
Temperature
 determinations, 66
 effects, 5, 191
Test methods
 chemical, 66
 concentration, 51–3
 dipping process, 183, 203–4
 latex thread, 235
 particle properties, 62–5
 polymer composition, 60–1
 volatile unsaturates, 61–2
Tetramethyl thiuram disulphide, 231
Thiazole, 230
Thickeners, 7, 79, 138, 158, 163
Titanium
 chelates, 159
 dioxide, 156–7, 165, 171
TMTD/ZnO/latex, 12, 14–16
Total solids content, 52
Tubing extrusion, 246–7
Tyre-cord dips, 250–2

Upholstery fabrics, 113–16

Vee grooving technique, 141

Vinyl
 acetate, 31–3, 123
  applications, 33
  residual determination, 62
 acetate–acrylate, 29, 113
 acetate copolymers, 116
 acetate–ethylene, 98
 acetate/ethylene/vinyl chloride latices, 165
 chloride monomer, 41
 pyridine latices, 38
Vinylidene chloride, 39–41
Viscosity determination, 62–3, 65, 183
Volatile fatty acid number, 14, 57
Volatile losses, 5–6
Volatile unsaturates, test methods, 61–2
Vulcanisation
 accelerators, 7
 glove manufacture, 196, 198, 199
 natural rubber latex, 19
 protectives, 188, 190
Vulcanising agents, 7

Wall cladding, adhesives, 132–3
Water-phase effects, 153–4
Water repellent, 247–8
'Wet picking', 96
Window blinds, polymer treatment, 116
Wood
 adhesives, 135–42
 veneering adhesives, 142
Woven fabrics, polymer treatment, 113–16

XSBR latex, 71–2, 75, 77, 78

Zinc
 ammine heat sensitive gelation system, 82
 dibutyldithiocarbamate, 230
 diethyldithiocarbamate, 230
 dimethyldithiocarbamate, 231
Zinc oxide viscosity (ZOV) test, 59
Zinc stability time (ZST) test, 14, 17, 59